Coping with Chaos
Seven Simple Tools

Coping with Chaos
Seven Simple Tools

By Glenda H. Eoyang

First Edition

Published by
Lagumo Corp.
1914 Thomes Avenue
Cheyenne, Wyoming 82001

Cover design by Lagumo Corp. The hummingbird photo-graph was taken by award-winning photographer and author Paul Wagner, and is used with his kind permission.

ISBN 1-878117-15-7

Acknowledgements

The creation of this book has been a complex adaptive process. Many experiences, opportunities, and persons have contributed to its emergence. My intent has always been that each person receive immediate and private thanks for support. For any I missed, thank you. For all, thank you, again. Some persons, however, made extraordinary contributions to my learning, and I offer them this public appreciation. The stories of Ralph Stacey and Frank Dubinskas enrich my learning as they do the pages of this book. My family and friends, who serve as teachers and mentors, provide the material on which this praxis of complexity is based. My business partner at Excel, Becky Bohan, helps establish the solid foundation from which I learn about chaos and complexity. John, the light of my life, nourishes me in body, mind and soul. I thank you all.

Glenda Holladay Eoyang
February, 1997

Table of Contents

Prologue
These Chaotic Times

Prologue
These Chaotic Times

Weird stuff is happening in the organizational world today. You read about it in the popular press; you hear about it at professional gatherings; you witness it in organizational meetings; and, as a manager, you live with it every day. At some point, you would think the weird would become common place, and everyone would stop talking about

- Too much work and too few hands to do it.
- Cross-cultural misunderstandings.
- Bizarre outcomes from inconsequential circumstances.
- Technology that evolves faster than it can be integrated.
- Budget and schedule changes that defy prediction.
- Criteria for success that change without warning.
- Corporate mergers or disintegrations that send tremors through sedate organizational structures.

In spite of evidence to the contrary, some managers continue to expect predictability in the midst of the weirdness. Others give up the search and are content to think of themselves as victims of insanity beyond their control. Neither of these "solutions" succeeds in turning the tide of weirdness. Neither of them brings a sense of security to the world of the manager. We need new strategies to work within a world where strangeness becomes stranger over time and where no explanation maintains its usefulness for long.

The Manager as Map Master

There was a time when it was helpful to think of the manager as a traveler in a civilized land. Every potential route was charted, and the manager's job was to select the map that matched the territory and to share it with employees. The job of the employee was to follow the prescribed trails and to arrive at their expected destinations on time and within budget. Those were the days of annual performance objectives and static job descriptions. If followers discovered the map they'd been given didn't match the real terrain, they simply returned to the leader's tent and requested a new map. The leader would then go to the map drawer to select a better guide or trot over to the leaders' leader's tent and request a new atlas. If the right map couldn't be found in the corporate library, a consultant was hired to provide a whole new set of maps. Everyone believed that someone, somewhere had the right map, and all we had to do was search until we found it. Things seemed weird to us because we didn't know enough, but someone must have traveled these same roads before and be willing to lend or sell us the perfect map.

At some point, this belief began to fade. Too often, we came to the conclusion that no one had the map we needed, and slowly we realized we were traveling in the wilderness. Well, that's okay, we thought. All we need to do is to scout out the territory and chart it. By creating new maps, we will extend our atlases and expand our libraries, then we will be working in civilization again. These were the days of mission and vision statements, benchmarking and Total Quality Management. We believed new maps could be created by the people closest to the work, so we empowered employees to merge their collective wisdom and build the map for the future. When the cartography was complete, we expected to be able to return to the old world of predictability and reliable planning and execution. Many of today's managers and management pundits continue to hold onto this hope: At some point, we will understand the lie of the land, and what we thought of as weirdness will just be another interesting turn in a well-traveled road.

Slowly but surely, this myth, too, is beginning to fade. Some managers in some organizations are realizing the problem is not the maps at all. No matter how many maps we have, and no matter how much detail we include in them, they will never be able to represent the reality in which we work. The source of our problems, the genesis of the weirdness, lies in the territory itself.

The world in which we travel is not only uncharted; it is unchartable. Like a shifting sand dune, it evolves from moment to moment in surprising and unpredictable ways. Any attempt to freeze and explain a current state is destined to fail because at the next moment a new reality will emerge, which might bear little resemblance to any previous state. Not only must we discard our old maps, but we must give up the vain activity of cartography. The role of the manager can no longer be keeping and distributing maps nor encouraging employees to design new maps.

Both the efforts are futile and wasteful, but what is an alternative? Must we resign ourselves to working in the dark, constantly feeling our ways through complicated topography until we are all hopelessly and helplessly lost? No, we can change our expectation of reality and search for other guiding tools and principles to allow us to make productive decisions in spite of overwhelming uncertainty.

I believe the sciences of chaos and complexity can define a set of *questions* to guide our actions when *answers* are completely beyond the bounds of prediction. Since my introduction to chaos, I have studied, written and spoken about the ideas. I have applied the concepts to my own management decisions and helped others use the metaphors to investigate and resolve their own organizational issues. The ideas presented in **Coping with Chaos** are a distillation of those learnings and teachings.

This is a pragmatic management book, it is not a book about science. At no point will I weigh you down with scientific detail or mathematical rigor. At some points, I will take liberties with the scientific concepts as I apply them metaphorically to organizational and management behavior.

The purpose of this book is to describe the strange behavior of complex systems and to investigate ways similar behaviors shape our organizations. We will explore the questions that are necessary to an understanding of complex systems and how we can use those questions to replace the need for maps and map making in tomorrow's organizations. From this perspective, weirdness is not the occasional aberration that requires controls, it is the innate power of the system to adapt to ever-changing environments, but we cannot throw off the paradigms of the past without replacing them with others more sensitive to the realities of the future.

The Old and the New

The old models of linear, Newtonian science and those of the new sciences of chaos and complexity generate management paradigms that are radically different. The table below summarizes the common, Newtonian paradigm first and then demonstrates how the complex systems paradigm stands in stark contrast.

Newtonian Perspective		Complex Perspective
Machine-like	◄─►	Organism-like
Linear	◄─►	Nonlinear
Predictable	◄─►	Surprising
Orderly	◄─►	Patterned
Controlled or Controlling	◄─►	Adaptable or adapting
Designed	◄─►	Emergent

The Newtonian concept of a corporation's functioning leads to some very specific actions on the part of managers.

- Management by objective. A competent, committed employee should be able to predict and control the environment to meet pre-determined goals and objectives.
- Individual accountability for performance or mistakes. The organization is a machine. If it is not working, then one or another of the component parts is broken and must be "fixed" before full functioning can be sustained.
- Preference for "yes men." Managers need to believe that their plans and predictions are correct. They show a preference for persons who contribute to the perception of omnipotence.
- Lack of respect for individual employees. The members of the organization are like cogs in a machine. Individuals' unique contributions cannot be encouraged or, in extreme cases, allowed at all.
- Secrecy regarding management decisions and plans. Cogs in a machine have no "need to know," so management controls access to significant information.
- Lack of innovation. New ideas threaten the stability of the leaders' world view. If the manager did not think of it first, the innovation will succumb to negative feedback or punishment.
- Undue attention to internal procedures and policies. Again, the concept of stability and predictability requires that management control all activities. One way to sustain control is by creating and enforcing detailed procedures.

These perceptions and the actions they engender, based on the Newtonian model, were functional at one time. They worked just fine when our markets were isolated from global markets, when our resources for expansion seemed unlimited, when government regulations established virtual monopolies in many industries, and when the work force was composed almost exclusively of white males. Today, however, when our business reality does not match those former assumptions, strategies based on the Newtonian, mechanistic model simply cannot work.

Another explanatory model exists. Based on recent research into the behavior of complex systems, this new model presents a world view in which systems, including organizations, are seen as organic wholes.

In the complex paradigm, one thinks of an organization not as a machine, but as a living organism. The role of the chaotic manager is less that of the driver or mechanic and more like that of the farmer. Farmers understand the forces that affect grain production. They provide resources, such as fertilizer and water, in a timely way to encourage maximum production. They coordinate planting and reaping so that all fields are harvested at the optimal time. They recognize that no two days and no two seasons are identical, so they continually watch the development of their crops and listen to weather and market reports to determine the most effective actions to take.

Chaotic managers are not shirking the responsibility of management, any more than a farmer would consider refusing to plant, till or reap. Their actions are taken in consonance with the natural development and individual contributions of each plant and each field. Effective action in this complex world requires both the structure and predictability of effective Newtonian techniques and the flexibility that comes from a chaotic understanding of the world and the role of the manager in that world.

The two paradigms—linear predictability and chaotic adaptation—are not mutually exclusive. You do not have to choose between them, selecting one and rejecting the other. Frequently, some aspects of a problem will be predictable and structured while others will be complex and surprising. Practice evaluating a situation and selecting the tools and techniques that will contribute to the productivity or effectiveness of your group. You can develop your skills in both paradigms and choose to behave according to one in some cases and the other in other cases. Think of the two paradigms as tools in your toolbox. In the same way that you would not use a hammer to tighten a screw, you should not use complex systems techniques and strategies to solve Newtonian problems.

Because the complex systems paradigm is new and unfamiliar, this book is dedicated to its explication. You can read more about the Newtonian paradigm in almost any other management book, and I encourage you to understand and use Newtonian techniques as well as chaotic ones. The problem, however, is to decide which approach is best for a given situation. The lists below describe situations in which each of the paradigms will be most effective.

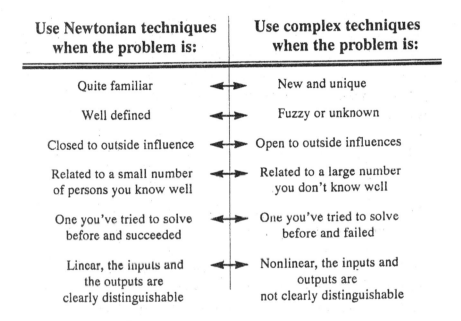

Use Newtonian techniques when the problem is:		Use complex techniques when the problem is:
Quite familiar	◄►	New and unique
Well defined	◄►	Fuzzy or unknown
Closed to outside influence	◄►	Open to outside influences
Related to a small number of persons you know well	◄►	Related to a large number you don't know well
One you've tried to solve before and succeeded	◄►	One you've tried to solve before and failed
Linear, the inputs and the outputs are clearly distinguishable	◄►	Nonlinear, the inputs and outputs are not clearly distinguishable

Mechanics of the Complex System

All of us have extensive education and experience with the Newtonian principles and their applications to management decision making, but those learnings cannot be generalized to

9

apply to complex systems. This section presents the special characteristics of chaotic systems that give them their surprising power for change and productivity. Like most complex adaptive concepts, it is easier to understand them within their own context, so we will introduce the tools of complexity with a story.

Blame It on Sally

I worked once with a small entrepreneurial company. It was an association of highly skilled technicians who individually provided excellent products and services to their clients. The company was based on the idea that all employees were equally talented, capable and committed to the common good. Everyone was included in evaluation of the cash flow, income statement and balance sheet. Marketing strategy and personnel decisions were always determined in open staff meetings. Even Mary, who was a natural leader and founder of the company, refused to acknowledge her superior insight and commitment to the corporate vision. We were all equal players.

During its third year of operation a major recession hit. All of the company's clients cut their budgets, and many refused to outsource work to companies like ours. At the same time, one of the most talented employees began to act strangely. The employee, Sally, suggested changes in marketing strategies and internal procedures. The rest of us rejected her suggestions out of hand because we were suspicious of her motives. She took all of her vacation time at once, ate lunch alone and kept her office door closed—clearly counter-culture behaviors. Because of the pressure on the market, the project load decreased, projects were smaller and had tighter deadlines, and all of us were slightly on edge. Sally's customers, especially, seemed to respond poorly to the bad economic times. Customers who historically had been responsible for 25% of our annual revenue were suddenly cancelling contracts and expecting extensive services without charge. All of us blamed Sally. We assumed her "strange" behavior was affecting the customers' confidence and loyalty to the company. When asked to report about progress on her projects, Sally became defensive and shared only minimal information. Her hesitation was read as sullenness, and we gave her multiple verbal and nonverbal messages about her "unacceptable" performance. We discounted her point of view and punished her for any difference or independence she expressed through her work or communications.

Over a period of three months, enough friction developed between her and the rest of the staff that she was asked to leave. We all believed, when Sally left, things would improve. Of course, they did not. Within a week, Joe, another of our star performers, began to withdraw from the team. The economy continued to slide, and soon Joe, too, was on the outs with many of his colleagues. Like Sally, Joe stopped socializing with the rest of the staff and began to hoard information about his projects and potential sales contacts. He stopped submitting his work for internal review, so customers received sub-standard products from him. Eventually, Joe, too, was forced out of the company.

Of course, none of us was conscious of this sequence of events. We saw no connection between Sally's unconventional behavior and Joe's sudden change in personality. In retrospect, however, it became clear to me that we were trying to identify a single cause of our distress—first Sally and then Joe. We thought that if we could remove the substandard part, our machine would function efficiently again. The fact that this strategy was ineffective escaped our notice. Eventually, this and other management myopias forced the dissolution of the company.

Behaviors of the Complex System

My experience with Sally and Joe and many other similar corporate mistakes indicates that actions based on the linear understanding of organizations do not solve problems. They merely move the problems around. As we continue to identify and implement isolated solutions, we continue to generate new problems.

To identify real solutions to real problems, we must consciously recognize that the environments in which we work are complex, nonlinear and chaotic. In the context of the new descriptive model, the meaning of *chaos* is different from its everyday usage. In general parlance, chaos means lack of order. It has been used historically to describe the vast emptiness that existed before the world was created, the crazy disordered action of individuals or groups and the random distribution of energy or matter throughout a system. In the context of this new scientific and management paradigm, *chaos* means something very different. A chaotic system may appear to be disordered or random, but it is not. Scientists have described complex systems in weather, the economy, physiology, ecology and thermodynamics. All of these systems share a few fundamental characteristics. When we understand those characteristics, we can begin to work with and within the complex forces that shape our corporate lives. The purpose of this book is to help you recognize these characteristics in your organization, understand the implications of these strange system behaviors, and adapt to your chaotic environment responsibly.

In Chapters 1 through 7 of this book, we will investigate the following characteristics of complex, adaptive systems:

Butterfly effects. In complex systems, a very small cause can have a tremendous effect. You would recognize a butterfly effect in business when a comment overheard at a water fountain results in a project's funding being halted. In the story about Sally, it is clear that her small changes in behavior resulted in profound changes in her status and role in the organization. As a manager, you can create and support systems that help positive butterfly effects flourish and that squelch the effects of negative butterflies. Chapter 1 provides information to help you manage butterfly effects.

Boundaries. Every complex system includes wide ranges of similarity and difference. The area that separates parts of the system that have different characteristics is called a boundary. Examples of boundaries in organizations include departmental distinctions, distinctions based on cultural backgrounds or gender, distinctions in personal style, and distance from the customer. Strange things happen at boundaries—information is distorted and behavior is misinterpreted. The boundary that developed between Sally and the rest of the staff made effective communication extremely difficult, if not impossible. The manager must establish and transcend boundaries to ensure that the right information reaches the right places at the right time. Chapter 2 provides information to help you manage boundaries in your organization.

Transforming feedback. Information flows across the boundaries in complex systems. This information generates changes in both the sender and receiver of the information. A change-causing exchange like this is called a transforming feedback loop. Feedback between Sally and the rest of the staff created destructive transformations for everyone concerned. Transforming feedback is the only control mechanism available in a complex system, so a manager must learn to use feedback effectively. Chapter 3 explains how you can use transforming feedback to shape the behavior of your complex system.

Fractals. A fractal is a geometrical object that results from multiple solutions of a nonlinear equation. The fractal provides a visual image for how a complex, chaotic system can encompass productive levels of differentiation and similarity from simple, core beliefs. When our organization responded to Sally, we were not able to balance differentiation and similarity to build a cohesive whole. Chapter 4 explains how fractals are generated and shows how the fractal can be a wonderful image of the complex, chaotic organization and your role as a manager in that organization.

Attractors. The interdependencies in complex systems appear to be random when viewed individually, but a global view of the system uncovers system-wide patterns of behavior. These patterns are called attractors. Sally's position in our corporation showed a two-point attractor that could not be resolved when we considered it in isolation. Managers can use the concept of an attractor to describe the over-arching behavior of a system and to take steps to influence the strongest, most productive attractors. Chapter 5 describes the types of attractors and explains how a manager can use the concept to build productive teamwork.

Self-organization. Complex systems exhibit another amazing property. When they become sufficiently disorganized, they generate their own order. None of us set out to exclude Sally and Joe from our team, but the powerful, nonlinear relationships among us caused the team's spontaneous split. This tendency, called self-organization, can help a manager generate order from disorder. Chapter 6 explains this amazing propensity and gives you tips about how to recognize and encourage self-organization in your business environment.

Coupling. Every complex system includes subsystems embedded in and related to other subsystems. Complex interdependencies knit these subsystems into comprehensible wholes. Coupling is the name given to the attachments between various subsystems within the complex system. Tight coupling among the other members in the team left no room for Sally and Joe, so they uncoupled from our corporate family. Managers can establish coupling that provides maximum freedom and maximum security for each subsystem involved. Chapter 7 defines three types of coupling and gives you tips about how you can identify and establish the most productive type of coupling between two interdependent subsystems.

The epilogue of the book presents a case study to demonstrate how complex, nonlinear relationships build both success and failure. The case describes a complex manufacturing project at Apple Computer. The engineering and management techniques and strategies are wonderful examples of how managers can establish and maintain projects that make the most of complexity.

Results of Complexity

As Sally's story demonstrates, organizations exhibit the characteristics of complex, chaotic systems. Though we must work within such complex structures, we must understand that their behavior is chaotic. Though it might appear to be random and uncontrolled, the complex dynamics within an organization can be understood in terms of the complex underlying structures and behaviors of complex systems. In the following chapters you will read about these seven characteristics of complex adaptive systems and learn how to use them to ask creative questions and take productive action in the midst of chaos.

Based on my learnings from hundreds of management professionals, it seems that many experienced managers have discovered chaos on their own. Many of my clients and students have said to me, "So that's what I've been doing, and that's why it works!" You, too, may have discovered the value of working constructively within chaos rather than trying to impose a Newtonian paradigm on a chaotic reality. If so, then the rest of this book will help you make those intuitions explicit. If not, then these simple tools for coping with complexity will provide you with a new approach to working productively in a world where maps and map making are a thing of the past.

Chapter 1
Butterfly Effects

Chapter 1
Butterfly Effects

Have you ever had a day disrupted by a bolt out of the blue? Have you been surprised to find your entire staff distressed (or ecstatic) about some comment you intended as a joke? Have you watched in amazement as a routine meeting flew off course and turned into a shouting match? Have friends or colleagues revealed that some casual comment of yours made a positive impact on their personal or professional development? If so, then you have already seen butterfly effects.

Simulations of weather systems demonstrate that if a butterfly flaps its wing in Tokyo, the tiny disturbance in the atmosphere might be magnified by other forces until it results in a hurricane off the coast of Florida.

This sounds like the beginning of a fairy tale, but it is not. In the strange and wondrous world of nonlinear dynamics, such remarkable things occur. Scientists call this phenomenon "extreme sensitivity to initial conditions." The description refers to the fact that complex, chaotic systems have the ability to magnify an extremely small variation in an initial condition. Through a series of nonlinear magnifications, a small cause can have tremendous effects.

What Are Butterfly Effects?

Many examples in the physical sciences demonstrate the existence of butterfly effects. The weather example, which gives butterfly effects their name, is a perfect demonstration of how complex systems are sensitive to initial conditions. In computer simulations, the butterfly wing in Tokyo really can cause a hurricane. These demons of predictability explain why long-range weather reports are doomed to failure.

Another example comes from the study of fluid dynamics. Imagine you are pumping water through a pipe. As you begin to pump, the water flows smoothly. You get a bit impatient, so you increase the pumping speed. The water is still flowing smoothly, but you want to finish the job even faster. So, you begin to increase the pump speed again, gradually. As the speed increases, so does the rate of the smooth flow of the water. At some point, though, turbulence develops. The water begins to flow backward in the pipe, and the efficiency of the flow is seriously eroded. You turn down the pump speed, and the smooth flow returns. You have identified a critical point in the pumping action. At one speed, the flow is smooth, but a very small change in speed results in extreme turbulence of the flow of water. The complex, nonlinear interactions among the water molecules and the microscopic irregularities of the pipe have created a system that is sensitive to the smallest change in water pressure. You have discovered a butterfly effect.

The classical paradox of time travel provides another example of the butterfly effect. This paradox makes a great basis for fictional plots. The paradox goes like this: If you could go back in time, you could observe some historical event, such as the moment your parents met. Imagine that during this little trip through time you stop your father to chat with him about his parenting philosophies. The chat is interesting, and you plan to broach the subject with him at your next family gathering. In the meantime, however, you have delayed him in his path. He walks into the bookstore two minutes after your future mother leaves, and the two of them have missed the opportunity to meet in front of the shelves of the used science fiction titles. Neither of them is the wiser as they meet and marry other people and have other children. But that leaves you unborn—unable to travel back in time to interfere in your father's travels. Because you are not there to interfere, your mother and father do meet, so you are happily conceived and born to travel back in time to interrupt your parents' meeting. And where does that leave you? This paradox demonstrates that today's reality is the result of an infinite number of complex, interdependent events. A miniscule change in any one of those events will result in a completely different future. Every little thing that happens might turn out to be a major influence on the future of the world. We cannot tell which factors will represent butterfly effects and which ones will cause only predictable changes. That is the nature of a complex, chaotic system.

All complex, chaotic systems exhibit this behavior. Because there are so many components and because those components interact in myriad ways, some behaviors will be amplified through time and others will be damped. Some individual events will change the subtle patterns for the whole system, and others will merely be lost in the system's underlying noise. We are left, then, with two questions: Do butterfly effects affect the behavior of complex organizations? What can a manager do to cope with a system that is extremely sensitive to initial conditions?

Take a moment to think about examples of butterfly effects from your personal and professional experiences. Have you ever been blind-sided by the effects of a decision you thought of as "not a big deal?" Has a customer, vendor or employee ever exhibited a strong reaction when you had no idea there was even a problem? Have you ever had to rework a complex project plan because an upper-level manager made an off-hand comment to your boss? If so, then you've seen evidence of butterfly effects in your organization.

Organizational Butterflies

If you have ever seen what you considered a tempest in a teapot, then you know the effects of butterflies in organizations. Some of the most common butterflies are related to the following situations.

Rumors. Did you ever play "telephone" as a kid? The game begins when a simple message is whispered from child to child. The player at the end of the line repeats out loud what he or she has heard. This simple, but distorted, feedback process is capable of transforming, "the blue balloon" into "ten rueful baboons." A little misunderstanding in any one of the transmissions can distort the message beyond recognition. As the distortion moves through subsequent tellings, it is magnified until the output of the process bears no resemblance to the input. This same butterfly is active in business rumors. A casual comment or humorous statement is converted into fact, and after a few more tellings is exaggerated into magnificent fiction.

Innovation. "Watson, come here, I want you." These were the first words transmitted over an electrical telephone. Alexander Graham Bell had spilled a beaker of liquid, the wires fell into the spreading pool . . . and the telephone was born. That day, Bell's slippery hands or stiff fingers became the butterflies that resulted in a revolution in communications technology. Many product innovations, such as the discovery of X-rays and penicillin, resulted from the same kind of accidental experience. The stage must be set for innovation, and researchers must be prepared to notice changes that are significant, but the immediate cause for success frequently is an unexpected and seemingly insignificant event.

The straw that broke the camel's back. Personal relationships also are victims of the effects of butterflies. A productive working relationship could be completely destroyed when one or the other participants makes a small misstep. A social slight, a misdirected memo, an unfortunate bit of body language caused by indigestion might be the straw that breaks the camel's back. The immediate cause might have no fundamental connection with the underlying problems in the relationship, but the butterfly can become the focus of attention. So, when the two try to reestablish their relationship, they resolve the "immediate cause" but fail to resolve the underlying issues. As a result, a minor infraction at a later date will again disturb the smooth interaction of the team.

I hate my job. Physical environments frequently introduce butterflies into the world of work. A friend of mine was miserable in her job. She left work each day with a headache. She assumed the work itself and her co-workers were stressing her beyond her ability to cope. One day, someone borrowed her computer to complete a rush project. While he worked at her desk, he turned on a hidden light, which illuminated the working area with a soft glow. When my friend returned to her desk, she left the light on—and went home without a headache. The problem had been eye strain! It was not long before her view of her work was positive and productive. The fact that she did not know about the light had blighted her physical comfort and personal productivity. The decrease in productivity worried my friend, and she blamed others for her distress. This small cause had tremendous effects—and was discovered only because of someone else's deadline!

Other types of information and organizational activities provide occasions for generating butterfly effects. The missing comma in a computer program that results in the loss of thousands of dollars; a casual conversation with a friend that includes a leak of critical competitive information; the loss and replacement of a staff member with someone who has identical skills but different work habits; a manager's thoughtless comment that results in extensive rework are all common examples of butterfly effects.

Positive or Negative—That Is the Question

While many of the common butterfly effects are destructive, butterflies can also bring about positive results. The mechanism for both is the same: A small change in a complex system is magnified out of proportion until it creates a tremendous effect, either constructive or destructive. The difference is that constructive butterflies move the complex system toward acknowledged productive goals, and negative butterflies move the system away from those goals. In the examples above, innovation and discovery of the hidden light demonstrate the potential for positive butterfly effects.

As a manager, one of your responsibilities is to interpret the impact of a specific butterfly effect and take action to magnify the good ones and squelch the bad ones. The tips in the next section will tell you how.

What Can I Do About Butterflies?

Recognizing butterfly effects is only the first step toward managing them appropriately. As a manager, you need specific strategies to help you cope with the small variations in your group that result in enormous effects. This section provides information about how to deal with the butterflies that can threaten or enrich your group.

Newtonian Approaches to Butterfly Effects

First, let's consider how a Newtonian manager would deal with a butterfly effect. These strategies are probably most familiar to you. You might have used them or been affected by another manager who used them.

Newtonian managers consider their groups to be machines. This implies a single cause for each single effect. When a butterfly appears, the manager assumes any major effect must have a big cause. The Newtonian solution is to identify the cause and fix it. So, when a butterfly erupts, the manager can spend a great deal of time searching for the cause. You've probably seen managers who conduct inquisitions to determine who, exactly, started a rumor. This approach, though firmly rooted in the Newtonian paradigm, is useless in a complex system. First, the single cause is so small, it might be impossible to identify. Second, even if you find the initial condition that resulted in the effect, there is no reasonable action you can take to undo the butterfly effects that resulted from it. By the time you are aware of a butterfly effect, many interactions have already embedded the butterfly information into the organization as a whole. The Newtonian approach is futile against butterfly effects that arise in complex systems.

Chaotic Approaches to Negative Butterfly Effects

In contrast, the manager of a complex system is not preoccupied with treating the butterfly effect at its cause. Instead, this manager watches for butterflies and seeks ways to stop the spread of negative and encourage the spread of positive butterflies. Further, the chaos manager establishes standard communication patterns that will provide data about butterflies before the effects become destructive, identifying the aspects of the system that allow butterflies to flourish. Responsive action, then, consists of limiting the effects of negative butterflies and making systemic changes that encourage appropriate butterflies.

To control the spread of negative butterflies you—as a chaotic manager—take specific steps like those described below.

Know one when you see one. How can you recognize a butterfly effect in your organization?

- You receive a message whose implications or interpretations do not logically follow from the facts presented to you.
- Something entirely unexpected happens.
- You notice strong emotional reactions to information that to you appears to be neutral.
- Several people come to you on the same day with different versions of the same story.

None of these, by itself, is sufficient to prove you are working with a butterfly effect, but any one of them can be a hint that the effects you see are results of butterfly wings.

Understand the flow of information across boundaries around and within the group. Boundary conditions, described in detail in Chapter 2, provide a wonderful mechanism for coping with butterfly effects. Your group is able to perform its work because it receives information from outside and distributes information within the group's boundaries. The flow of information can be formal or informal. It is the informal flow of information that is fertile ground for butterfly effects. Know where your people get their information and how the information tends to flow across boundaries within groups. Notice who talks with whom and who likes to retell a good story. Recognize that these communication dynamics will affect the transmission of potential butterfly effects. Reinforce communication paths that provide useful and constructive information and ignore those that do not.

Be careful not to punish persons for their communication networks, because punishment is, frequently, a powerful reinforcer. It focuses attention on the very thing you want to de-emphasize.

Provide ample information in a timely fashion. Butterflies are most frequent when good, reliable information is not available. Your staff members want to know the facts and your interpretation of the facts. If you do not fill this need, then someone else will, and the information from that source might not be conducive to the group's on-going success.

Refuse to participate in the process that exaggerates negative butterflies. As a manager, you also participate in the complex interactions of your complex system. Like all other employees, you might be responsible for exaggerating the effects of a negative butterfly. That is why it is so important to recognize one when you hear it. Whatever you do when you suspect a butterfly, do not overreact to the content. If you begin a thorough investigation or demand the stories stop, you will merely escalate the perceived importance of the butterfly effect and prolong its life.

Use humor to defuse the emotional effects of butterflies in an organization. Humor is a wonderful tool for relieving stress. As a manager, your laughter can send a strong message that the content of a negative butterfly is absurd. Remember not to direct humor at an individual person. Rather, demonstrate that you think the situation itself is unbelievable and ridiculous.

Give it time to diffuse itself. Many butterfly effects blow themselves out in a relatively short time. If you can ignore the effect, then wait and watch. If no one is continuing to feed the cycle with new information or new responses to the information, then people lose interest and return to more satisfying endeavors.

As a manager, you will hope those more satisfying endeavors include a good day's work!

Chaotic Approaches to Positive Butterfly Effects

As a manager, your other main butterfly responsibility is to find ways to encourage positive butterfly effects. They can be the source of innovation, creativity, improved confidence and energy or just plain fun. You can take a variety of steps to stimulate the generation of positive butterfly effects.

Establish communication networks within your group. Use your knowledge of who talks to whom to build a mental model of informal information flow in your group. You might even wish to diagram the flow of information to help you know where and how a positive butterfly might move from one person or team to another. Establish and participate in some of the informal communication paths so you become an accepted part of the complex information network that can produce and magnify butterfly effects.

Stay in touch with people outside your group. If your group is isolated from the outside world, you will find few opportunities to identify positive butterflies. You can maintain communication with your peers within the organization or with friends and colleagues outside the corporation. Continually search for information that might prove useful to the work of your group. Introduce those ideas and help magnify them into productive butterflies.

Identify small bits of positive information that might produce butterfly effects. Keep your eyes and ears open. Identify any positive thing, even if it is quite small. If it is new, relevant to work or particularly interesting to your staff members, it might generate sufficient momentum to produce butterfly effects. Don't forget to watch for accidents. Often, an accidental discovery can be amplified into a powerful productive force.

Use your interpretive and distribution skills to magnify positive butterflies. After you have identified something that might prove to be a positive butterfly, focus on it. Describe it to your staff in terms of its productive possibilities. Work within the communication networks you have established to bring attention to the new idea or opportunity. Each of your interactions will be one of the transformations that moves a small cause toward a large effect.

Talk about the history of the group and incorporate new, positive events into the living history. In a complex system, history cannot be repeated, but it does build a framework for the future. As a manager, you can collect bits of information that might appear meaningless at the time, interpret them to represent a constructive characteristic of your group and share them with others. This progressive process allows you to define a self-awareness that magnifies the effects of a single, seemingly insignificant event.

Tell stories. Story and myth is the stuff of positive butterflies. Select an appropriate image that reflects the best in the group, and plant that image in multiple communications. Tell the story and encourage others to do the same. After the seed is planted, watch it grow and spread across the organization.

Keep your ear to the ground. The butterfly and its effects live in the day-to-day communications of a group. Take every chance to tap into the rich sources of information that come from informal conversations with and among members of the team. Stay sensitive to news that might shift the communication networks or introduce valuable information into the working environment.

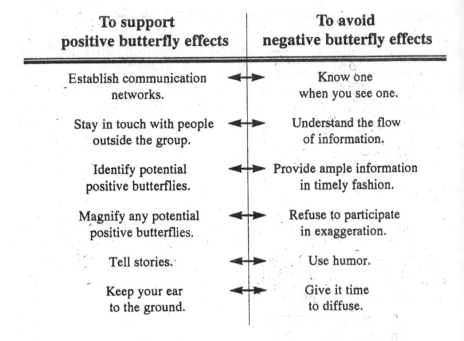

To support positive butterfly effects	To avoid negative butterfly effects
Establish communication networks.	Know one when you see one.
Stay in touch with people outside the group.	Understand the flow of information.
Identify potential positive butterflies.	Provide ample information in timely fashion.
Magnify any potential positive butterflies.	Refuse to participate in exaggeration.
Tell stories.	Use humor.
Keep your ear to the ground.	Give it time to diffuse.

All of these tactics can help you make the most of the butterfly effects that bring surprises into your everyday life.

Butterfly Effects: Two Examples

I have had frequent occasions to deal with the effects that lie in the wake of butterflies. The descriptions of two of those examples follow.

Public Relations

PR professionals have perfected the art of butterfly management. They take a small piece of positive information and find ways to amplify that information through the press and other public communication media. They take a small piece of negative information and embed it in a bundle of positive messages. The hope, of course, is that the positive butterflies preoccupy the public until the negative message has been forgotten. When McNeil discovered that someone had added poison to their Tylenol capsules, it immediately publicized its efforts to remove the product from shelves, redesign the packaging, redefine the product and support the investigative efforts of police. Sure enough, the public conversation turned from the tragedy to the valiant efforts of McNeil. The experience was costly for the company, but Tylenol is still on the market today.

Technology and Butterflies

Because technology shortens the time required for one interaction in a system, and because it makes interactions more frequent, it tends to multiply the occurrence of butterfly effects. Fax machines, phone mail, e-mail, radio and television all increase the speed and power of butterfly effects in complex human systems. One of my staff members experienced such a technology-generated butterfly effect. She was working on a project that was showing signs of difficulty. It appeared a deadline would be missed, so she documented the causes for the delay and sent an e-mail note to the primary customer. Well, she was working on someone else's terminal and did not realize that the SAVE and SEND TO ALL buttons were so close together. As soon as she pressed the button, she realized she accidentally had sent the detailed, unfavorable project description to everyone in the division. After a moment of panic, she sent another note to everyone, asking each recipient to delete the preceding message and explaining her mistake. She then went to see the primary customer and told her about what had happened. Together they decided to wait and see what would happen. The only effect from this butterfly turned out to be some pretty rough teasing about computer literacy and a slight redesign of the graphical user interface.

This chapter has introduced you to the origin and impact of butterfly effects in complex, chaotic organizations. As a manager, one of your tasks is to manage the kinds and strengths of butterflies generated in your group. You can use the tips in this chapter to help you design both proactive and reactive strategies to cope effectively with butterfly effects in your complex organizational world.

Chapter 2
Boundaries

Chapter 2
Boundaries

A complex system thrives on difference: Differences in air pressure generate the wind, differences in temperature set the stage for convection currents in liquids, a difference in ion concentration osmotically moves life-giving nutrients into cells. In a complex, highly interdependent system, variations in the system provide opportunities for movement and change. And the nature of the boundaries within the system determine whether the change is gradual or violent.

What Is a Boundary?

A boundary is the area that lies between two different parts of a system. The conditions at the boundary determine the nature of the changes resulting from the differences.

Organizations, like other complex systems, require differences for change and adaptation to take place. Variations in organizational levels, opinions, business functions, cultural contexts, education and gender all present opportunities for transfer of information, energy or materials among parts of the corporation. This transfer and the change it engenders, is what we recognize as work. If you wonder why you spend so much time in meetings, consider this: The meeting is the opportunity to identify and respond to differences that allow your institutional system to adjust. They are the occasions for organizational winds to blow, for the currents of change to be established and for life-giving information to flow osmotically from one individual contributor or team to another.

Transfers between the same and the different are necessary to business activity, but they are not always gentle or constructive. Frequently the differences in an organization generate bias, frustration or anger. Rather than a gentle breeze, the differentiation can result in thunderstorms and tornadoes. Boundary conditions are the characteristic of the system that determine or describe how two different parts of a system relate to each other.

Kinds of Boundaries

The many different kinds of boundaries that occur in nature can be seen metaphorically in organizations. In this section, you will read explanations and examples about four different types of boundaries.

- Clear and distinct
- Indistinct
- Partially permeable
- Impermeable

Clear and distinct boundaries. Sometimes, the boundary is clear and distinct. In a weather system, for example, the boundary is called a front—cold, dry air on one side and warm, moist air on the other. Between them lies the boundary where the differences meet each other head-on. Clear and distinct boundaries usually result in turbulent change, such as turf battles and jealousy. We see examples of these clear-cut boundaries in organizations when union and management sit down at the bargaining table, when new technology is introduced into an office, when a manager confronts an employee who has been caught stealing, when a new and aggressive competitor enters the market. In each of these cases, the change promises to be traumatic, though the outcome can be seen as either constructive or destructive.

Jack James and Sam Smith started Snapit Industries, a small manufacturing firm, in 1990. When Jack and Sam began the business, they thought of each contract negotiation as a clear and distinct boundary. The customers had one vision of products, services, time line and cost; Jack and Sam had different expectations. Each negotiation over contract details became an "us against them" encounter. Jack and Sam prepared for the meetings as if the differences between them and their clients were more important than the similarities. Negotiations were painful and traumatic, and projects were always well underway before the relationship with a primary client would reach even a semblance of teamwork. Jack and Sam ultimately realized how they and their clients were behaving, and they decided it was time to approach boundaries in a different way.

Indistinct boundaries. Sometimes, boundaries are indistinct and difficult to recognize, such as those in a pot of heating water. You can know for certain the water at the bottom of the pot is hotter than the water at the top, but there is no distinct line differentiating hot from cold. Because the boundaries are not obvious, they cannot be viewed directly. One must observe the behavior of the whole system to understand their presence and then infer their influence on the system. For example, you have no visible evidence of temperature differentiation in the pot until you see the wavy convection currents, though heat might have been applied to system for quite some time.

Such boundaries are the most difficult ones to deal with in organizations. Their effects are subtle and delayed in time. A difference might grow within a system for an extended period, without showing any visible evidence. Then, suddenly, strange behaviors emerge without any apparent cause. Examples include: Personality conflicts, differences in working style, variations in perceptions of the corporate vision, nepotism or "good old boy" networks, unresolved resource battles and inconsistent hiring practices across an organization. These conditions all introduce differences that can have profound effects on the the functioning of a corporation, but because their boundaries are unclear, specific, focused action might not be possible.

In their third year of business, Jack and Sam entered into a joint venture with one of their clients. Each side shared a strong mutual respect for the other and both had an innovative product idea. On that basis, they drew up an agreement.

All went well for the first three months while they talked about product design, but in month four, tensions began to rise. Distrust and anger grew among the individuals and between their companies until all work stopped and the project was placed on long-term hold.

What was the underlying problem? The agreement for a joint venture disguised the historical distinction between customer and vendor. During design, no problems appeared because all were accustomed to working as peers through that part of the process. But the customer had always had complete control over the production and marketing of previous products, so when the new project entered those phases, historical distinctions and lines of control influenced the working relationship. At the time, the parties were not aware of the boundary conditions. They just knew that without a doubt they would never work together again.

This frequently is the outcome when boundaries are not made explicit and dealt with through an understanding of complex system behavior.

Partially permeable boundaries. Sometimes, boundaries are only partially permeable. Cell walls, for example, will restrict the flow of some chemicals and allow the flow of others. Such a boundary is more complex and requires general rules to define what passes and what does not. These boundaries allow selective adaptation of the system and less traumatic movement within it. We see such boundaries in organizations when individual and corporate confidentiality are wisely observed, when social and business relationships are differentiated, when middle management filters information flowing from employees to upper management or from executives to employees, and when "need to know" drives the dissemination of information.

As Jack and Sam recognized the need to change their negotiating tactics, they moved toward establishing a semi-permeable boundary between their company and their clients. Each negotiating session began with their efforts to understand the client's financial and schedule constraints. This effort was followed by a short presentation in which Jack and Sam explained their estimating procedures and manufacturing and distribution processes. On this foundation, discussion of the contract details progressed smoothly. The difference between vendor and customer was still clear, but selected information flowing across the boundary had eliminated the need for traumatic interaction. The customer relationship was not damaged, and Sam and Jack were surprised to find their bottom lines improved, as well.

Impermeable boundaries. Some boundaries are impermeable—nothing is allowed to pass between the different parts of the system. The process of natural change is aborted by the presence of boundaries that are impassable barriers. While humans have a propensity for building impermeable boundaries, examples in nature are rare. Interdependence and permeable boundaries have encouraged adaptation and evolution of natural functions, while impermeable ones have been minimized through natural selection. One example of an impermeable natural boundary is the impossibility of procreation between species. This impermeable boundary has survived in almost all natural situations, leading to clear differentiation among groups of living things. Plant hybridization and mule husbandry are clear exceptions, in which cross-fertilization produces generations with new and useful characteristics. Impermeable boundaries freeze a system in a current state and disallow interaction with other, different parts of a system. Examples of impermeable boundaries abound in today's organizations. They include hierarchical organization charts, confidential appraisals of individuals' performances, geographical distances, building security systems, intellectual class structures, social cliques and inaccessible computer data. All of these evolve or are designed to disallow adaptive change so that a difference is sustained in the system.

One of Jack and Sam's major clients was undergoing enormous organizational change. Over a period of one year, the client's company had reorganized three times. Each reorganization disrupted Jack and Sam's projects, distressed their employees and forced significant changes in internal processes in their company. During a heart-to-heart talk with their most recent primary contact, Jack asked if the worst was over. The answer was the one he dreaded: The changes would continue. In fact, a new CEO probably would be hired within the next month, and no one could guess what changes she would institute. As he drove back to the office, Jack hatched a plan to cope with the customer's upheaval. When Sam agreed, they announced the plan to the staff.

They decided to stop reacting directly to every change. Employees were instructed not to talk with the customer representatives about the organizational changes and not to talk with each other about the implications of those changes. The staff was assured that, until things settled down, Jack and Sam would require no more internal process improvements based on the customer's organizational structure. Of course they would continue to provide goods and services, but they would build an impermeable barrier between themselves and the customer until relatively permanent changes were defined. From the employees' point of view, the boundary was totally impermeable—changes in the client company would not affect their own. Both Jack and Sam maintained communications with their client so they would know when it was safe to replace the impermeable boundary with a permeable one again. This strategy decreased stress and increased productivity for everyone in the company. In the long run, it allowed Jack and Sam to provide appropriate services to their client during a trying time, without complicating the situation more than necessary.

Boundary Effects

Differentiation within a complex system allows for adaptation, and boundaries mark the interfaces across which the changes can take place. Boundaries in an organization can evolve naturally or they can be consciously designed to be distinct, fuzzy, permeable or impermeable, based on the types of differences involved and the desired rate and turbulence of change.

Boundaries and the conditions that surround them can be either destructive or constructive. The changes that emerge at the boundaries are like all change, painful or pleasant, encouraging or discouraging, generative or degenerative. A cold front might bring the gentle rains of spring or the fearsome destruction of tornadoes. In organizations, we can see examples of destructive boundary conditions in examples of bigotry, suspicion and hatred. We can also see constructive boundary conditions in training situations, collective decision making, dedicated customer service and productive partnerships. Only the context and immediate conditions determine whether the outcome will induce growth or death. That is why the manager of a complex organization has a responsibility to pay attention to boundaries and the conditions they engender.

Though you can think about boundaries and boundary effects according to the types described in the last section, in the real world things are not that easy. No complex system is based on a single boundary and the simple set of differences it divides. Because of the complicated interdependencies and the indefinite number of parts in a complex, chaotic system, you can expect differences across an infinite number of dimensions and an infinite variety of degrees of difference for each one. The boundaries overlap and distort each other. They shift unexpectedly, and they appear and disappear without warning. When you look at a system in all of its dynamic glory, the boundaries that affect its behavior usually are not obvious.

Boundaries in Teams

Consider the example of a team meeting. Among any group of persons, there are many kinds of differences and therefore many potential boundaries. Culture, education, gender, personal interests, friendships, experience, accumulated information, formal leadership responsibilities, professional history, personality styles and immediate preoccupations represent only a few of the boundaries that might affect the work of the team. Specific team activities define another set of boundaries dividing one team from other groups within the organization.

All of these differences and their correlated boundary conditions continually are present in the work group. At a particular moment, however, one or another might dominate the group interaction. For example, if someone tells a sexist joke, then the group might line up along the gender boundary, and constructive communication between men and women in the room might be jeopardized for some time. On the other hand, if one member addresses the group with arcane academic language, then the boundary between educational levels and disciplines will rule the dynamics until another boundary draws the groups' attention in a different direction. Of course, the most productive boundaries are the ones related to the formal assignment of the group, and an efficient group will focus its attention on those boundaries most of the time. Efficient use of boundaries seldom occurs by chance. All members of the team must minimize their irrelevant boundary conditions and optimize the productive boundary conditions for work to progress efficiently.

What Should I Look For?

How can you "know one when you see it?" Look for the following tell-tale signs of boundary conditions.

- Definable differences among persons or situations
- Information that is consistently distorted as it passes from one person or group to another
- Perceived or described tensions or other emotional reactions
- Marked change in behavior, based on circumstances
- Important topics that are discussed infrequently
- Extreme personal behaviors, such as withdrawal or aggression
- Frequent jokes about individuals or groups

Any of these signs might indicate a boundary condition is affecting the behavior of the group. If you find evidence of these characteristics, you then must determine which of the myriad number of potential boundaries is generating the behavior you notice and take action to shift the attention of the group to a different boundary.

As you read the following stories, think about what boundary conditions are in effect and how they distort the behavior of the system. The explanation that accompanies each story describes the situation according to its boundary conditions.

The self-managed team. Mary, John and Jeff were shocked when their supervisor, Betty, was laid off. The strangest thing was they were expected to report to Betty's boss, whom none of them had ever met. Betty had always taken care of them

43

and protected them from what she described as his, "fits of fancy." The day after the layoff announcement, Big Boss called them into his office and announced that from that day forward they would be a self-managed team. He said he had a great deal of respect for them and assumed they would rise to the task.

They left the meeting in a daze, determined to meet his expectations. Their first meeting as a team was an unmitigated disaster. Mary, who had been a close friend of Betty's, came to the meeting with an outline of tasks to be assigned to each person. John, the quiet one, looked up from his newspaper long enough to insult Mary and explain that he knew what he needed to do and should just be left alone. Jeff suggested they attend a training session to develop the interpersonal skills they needed to work as a team. They all left the meeting frustrated and angry. During the rest of the day, they appeared one by one in Big Boss's office to complain about the meeting and seek further guidance about how to proceed.

What boundaries did you identify? Consider the following ones.

- The emotion-laden difference between employees who are laid off and those who remain
- The impermeable boundary Betty had established between her staff and her boss
- The gaping, semi-permeable boundary between Big Boss and the team
- The distinction between what Big Boss thought they could handle and what they were trained to do
- The different perspectives about what the term "self-managed" implied
- The boundary, based on friendship, separating Mary from the other team members
- Distinctions in personal preferences and working styles that coalesced in the divergent strategies presented in the first team meeting
- The lack of loyalty and mutual commitment forcing members to seek resolution with Big Boss, rather than with each other

While knowing which boundaries are at work does not necessarily tell you what should be done to rectify the situation, it does allow you to ask productive questions as you consider possible corrective actions.

Users and information systems (IS) professionals. Everyone was excited. The new project was going to be a great! It included new client server technology, new graphical user interface design and re-engineered business functions. Senior management in IS and user communities had written the business case, defined the project scope, completed the preliminary estimates and gotten approval for initial schedules and budgets. The pressure was so great to get the project started that members of the project team were not consulted on any of these basic matters.

The project kick-off meeting brought together representative users, who would provide content expertise, and the technical team that would complete the formal analysis and system-design functions. It was a wonderfully productive meeting, and everyone was certain this project would forever change the way business software was developed. Three weeks later, the project was beginning to turn sour. The user representatives had to cancel meetings because daily work took precedence over the project. Analysts produced flow charts and data flow diagrams users couldn't understand. First line managers in IS were given authority to assign tasks and resources, while user supervisors had to get approval for every decision. Hardware and software constraints were mysterious barriers for users and points of endless fascination for programmers.

In week four, management on both sides of the fence met to evaluate the progress of the project. They spent hours listing and discussing the variety of problems and left the meeting without a clue about how to resolve any of the issues.

What boundaries do you see at work in this situation?
Consider the following ones.

- Everyone perceived the project as completely different from any attempted before.
- Planning was done by high-level management, without input from project leaders or individual contributors.
- Roles and responsibilities clearly divided the users from the IS professionals.
- For users, daily work competed for attention with project activities.
- Technical language and communication tools were inaccessible to users.
- Standard organizational structures and spans of control in the business group did not parallel those in IS.
- Intricacies of hardware and software capabilities and differences were significant to IS, but irrelevant to users.
- Accountability for analyzing and resolving the problems was allocated to upper management, though the problems were generated by complicated boundary conditions within the team itself.

No wonder the problem appeared to be insurmountable. Multiple boundaries and variable boundary effects were distorting the work of the team and, until the boundaries were defined and the effects mitigated, the team could not be either effective or efficient.

What Can I Do About Boundaries?

While identifying boundary conditions is an important step in problem analysis, simply knowing about the boundaries is not enough. A responsible manager must take action, when necessary, to alleviate the confusion, tension and counterproductive activities resulting from destructive boundary conditions.

The Steps

Here are some pragmatic steps you can take to influence boundary conditions and their effects on an organizational group. These steps might seem simple, but they open up a whole new world of options for management intervention.

1. Brainstorm about the possible boundaries affecting the group's behavior.
2. From that list, select the one boundary that seems to be most influential.
3. For the selected boundary, describe, in as much detail as you can, the apparent differences across the boundary.
4. Identify how the character of the boundary needs to change.
5. Brainstorm for tactics that will change the nature of the boundary.
6. Select one tactic to try first.
7. Try the tactic and collect information about the resulting changes.
8. Depending on the outcome, return to either step 2 or step 6.

The explanations below will help you take these productive steps.

1. Brainstorm about the possible boundaries affecting the group's behavior. This brainstorming activity will provide many benefits if it is completed by the group itself. First, it helps the group become conscious of its own dynamics in a non-threatening environment. Second, it allows individuals to voice their own concerns regarding their differences from other members of the group. Third, it acknowledges that differences are essential to creative group activity. Fourth, it can be fun, because identifying silly differences (such as tall and short or smokers and non-smokers) will lighten the atmosphere. This is a powerful activity. I have witnessed cases in which the boundary issues for a group disappeared as soon as this one step was completed.

2. From the list, select the one boundary that seems to be most influential. In this evaluative process, group members discuss the relative significance of the boundaries they have identified. Frequently, the long, brainstormed list will be consolidated into a smaller number of boundaries, as participants recognize that different behaviors arise from the same boundary. Differences of opinion about relative importance can also provide additional information about how the boundaries are affecting work on a day-to-day basis. Because this process can be repeated as many times as necessary, it is not critical the group select only one boundary for further investigation. They just need to pick the one they will work on first.

3. For the selected boundary, describe, in as much detail as you can, the apparent differences across the boundary. This step requires a series of questions. Frequently it is helpful to have group members answer these questions individually, and then to share their insights with the group as a whole. Questions to be answered here include

- Which side of the boundary are you on?
- If you move from one side of the boundary to the other, when and why do you switch?
- Who in the group is on which side of the boundary?
- How would you describe the characteristics of the groups on either side?
- How would each side of the boundary describe the other group in uncharitable terms?
- How would each side of the boundary describe the other group in charitable terms?
- When does this boundary interfere with the work of the group?
- When does this boundary improve or enrich the work of the group?
- What are the forces that generated the differences in the system that appear around this boundary?
- How distinct is this boundary?
- How differently is this boundary seen by different individuals?
- Currently, is the boundary permeable or impermeable?

These and any other relevant questions should be investigated by individuals and discussed by the group. Some benefit will accrue when one member or manager completes this process in private, but the interaction of the group is far more effective for permanent and constructive change.

4. Identify how the character of the boundary needs to change. In step 3, the group defined the current situation. In this step, they define a preferred future state. The discussion can be based directly on the discussion of step 3. Questions to be addressed might include

- Is is possible for one person to move from one side of this boundary to another?
- To meet personal or group needs, should you stay on the side of the boundary where you have been?
- In what circumstances should you move from one side to the other?
- How can we make change across the boundary more productive?
- How should the groups on each side of the boundary change to accommodate the group's needs?
- How must the groups on each side of the boundary stay the same to accommodate the group needs?
- How might this boundary be used to improve or enrich the work of the group better?

These questions will set the objectives for the change tactics you will determine in step 5.

5. Brainstorm for tactics that will change the nature of the boundary. Given a solid group understanding of the boundary and the conditions attending it, the group is prepared to investigate tactics that could make the boundary more productive. Some tactics work well to change boundaries.

- Holding more or less frequent meetings
- Developing listening skills
- Establishing group norms for behavior inside and outside meetings
- Turning focus away from this boundary and toward another one
- Acknowledging behavior pertaining to the boundary with verbal cues (such as, "I know this is an old-timer's view. . . .") or with silly physical cues (donning a red hat when making a statement reflecting the marketing, rather than technical, perspective)
- Providing training to erase differences in knowledge base or expertise
- Pairing people across the boundary in mentor-type relationships
- Allowing the two bounded groups to meet separately and come to a group meeting with joint proposals
- Identifying persons who sit on the boundary and relying on them to give observations of behaviors around the boundary

This is not an exhaustive list, and all of these tactics will not work in every situation. The nature of the difference, the nature of the boundary, the context of the group and the extent of their common tasks all contribute to the dynamics that built the boundary. These same factors are the ones that will determine which tactic or combination of tactics will be most effective in making the boundary more productive.

6. Select one tactic to try first. Do not try to implement more than one of the tactics at once. Group members will be overwhelmed, and focus will be dispersed. Not only are multiple approaches not practical, frequently they are not necessary. Boundaries in complex, dynamic systems are prone to change anyway. Since they are extremely sensitive to the contexts in which they appear, any change in the context often alters the boundary beyond recognition.

Be sure the tactic you select is one that can be adopted by everyone in the group, and recognize all who implement it. A self-reinforcing effect occurs when everyone in a complex system does the same thing. This reinforcement amplifies the effect by orders of magnitude, so it will reduce the amount of "brute force" required to implement change.

7. Try the tactic and collect information about the resulting changes. Chances are this process and its explicit attention to a boundary will already have solved the problem. Still you should collect data and confirm the results. Set a specific period of time—a day or a week. At the end of the time, check with the group to see how the boundary conditions have changed. Return to the answers generated in steps 3 and 4 and identify any differences in perception that have occurred since the original discussion. Evaluate the effectiveness of work around this boundary and discuss whether it requires further attention. If change has occurred, but more substantial change is required, you might decide to extend the experimentation period. You might decide the boundary is no longer an issue and turn to another one.

8. Depending on the outcome, return to either step 2 or step 6. If the boundary of focus is no longer a problem, you can return to step 2 to identify another boundary requiring attention. If the chosen tactic is not working, you might return to step 6 and select a different one from your brainstormed list. In either case, you can continue to iterate this process until the boundaries that interfered with productive work have been converted into areas of constructive, rather than destructive, difference.

Busting Boundaries: An Example

Consider the case of the software design project described earlier. The project suffered from many different boundary conditions between the user community and the IS professionals. After the executives met to discuss the problems, they returned to their project leaders and laid down the law: You will solve all of these problems, or we will take you off the project. So, out of desperation, they tried the boundary busting steps at the next team meeting.

Step 1. The group brainstormed their differences and boundaries and identified a wide range of differences. Some seemed relevant to their purposes and others did not.

Step 2. When the brainstorming slowed to a stop, the group reviewed the list. They marked out items that were not relevant, and they condensed other items into logical groups. When selecting the most important boundary, the conversation revolved around differences in knowledge—of the business and of the technical language and tools. The group agreed these two were primary and must be worked on. But which should come first? Looking at the other boundaries, they determined the users already had two jobs—the project and their day-to-day responsibilities. It made sense that the technical folks should take on the additional responsibility of gaining a new knowledge base. As a result of this discussion, the boundary they chose to address was the one around knowledge of the business.

Step 3. As they focused in more depth on the selected boundary, they came up with a variety of insights about what each of the groups knew and did not know about the business and the technical environment. Besides providing specific grounds for action, these conversations helped each side build new mental models of their work and their own roles on the team.

Step 4. The group considered the current boundary and identified the ways in which it was interrupting the free flow of information and energy across the group.

Step 5. With this understanding of the boundary, the group brainstormed for tactics that might work. They generated a brainstormed list of actions they might take to improve the understanding and knowledge base of the group.

Step 6. Based on this list, the group evaluated each option, looking for one that would be fastest and easiest to implement and would promise the most "bang for the buck." They selected the change in norms that would give everyone permission to ask for definitions of terms when they needed them. They decided to try this tactic for a week to see how it worked.

Step 7. One week later, they checked in to see how the new norm was affecting work. Initially, people had been irritated by the interruptions, but they lived with it in the hopes it would help. As the week passed, everyone noticed fewer people asked for definitions and upon reflection they realized two things had happened. First, the basic terms had been learned by everyone—this effect was expected. Second, the business experts began to speak more slowly and to define complex terms as they went along, reducing the need for interruption. Based on the questions they had been asked, the business experts had developed a fundamental understanding of what the others knew already and where help was needed.

Step 8. The business knowledge boundary seemed to be resolved, so the group was ready to address another of their list of boundaries. They returned to step 2 and, to their amazement, they discovered some of the other boundaries were less troublesome than before. This one small change (permission to ask for definitions) at a critical boundary had unexpectedly resolved tensions around numerous other boundaries. Because the business experts were talking more slowly and showing their concern for others' understanding, the technical people were more sensitive in their own speech. The increased understanding of the business and business process helped the technical teams remember to schedule meetings at times that were convenient for the users, who then could attend more regularly. Communications documents contained less technical jargon and began to look like recognizable pictures with English-language explanations. With the blessing of users' supervisors, IS team leaders learned to work directly with higher-level managers in the users' organization to resolve resourcing and scheduling issues, understanding the supervisors had neither authority nor time for such decisions.

Thus, one small change at a contentious boundary had solved a variety of seemingly unrelated problems.

Of course, I cannot guarantee this process will magically resolve all of the issues a group faces, but I have seen it work often enough to say that an understanding of boundary conditions provides a view into a groups' dynamics that is inaccessible from any other vantage point. And any process making the group conscious of boundaries will generate innovative and powerful solutions to complex and seemingly intractable problems.

This chapter has presented a process for identifying and coping with boundaries in organizations. It has presented questions you can ask yourself and others to uncover boundary issues that distort communication and distract attention from constructive work.

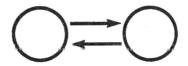

Chapter 3
Transforming Feedback Loops

Chapter 3
Transforming Feedback Loops

A complex system consists of an indefinite number of individual, interdependent parts. You can think of the parts as single persons who contribute to the work of a team, as teams working together for the good of a department, as departments that fulfill their various functions to meet the needs of an institution, as groups of companies that form an industry, or groups of industries that form the world-wide economic network. Any of these levels of organization can be thought of as a complex system. This chapter describes a special kind of interdependency that links the parts of a complex system into a functioning whole. We call the special attachment a transforming feedback loop.

What Is a Transforming Feedback Loop?

A transforming feedback loop is the form of communication that passes across the boundary between any two parts of a complex system. One part of the system changes, and the change is transmitted to another part of the system. The second part responds to the change in the first, and its change is transmitted back to the first, which responds with a change of its own. Such a transfer and the changes engendered by the transfer constitute the transforming feedback loop. This transaction permanently alters both the sender and the receiver.

Nature gives ample evidence of the importance of transforming feedback loops. For example, the air over a large land mass is warmer than air over a mass of water. As it is warmed, the air over the land rises and cold air from the water rushes in to fill the low-pressure area generated by the rising warm air. The new, cooler air is warmed and it, too, begins to rise. When the air rises high enough, it begins to cool and to fall toward the earth. It cannot fall toward the land mass because the warmed air is rising from there, so it must fall over the water. This, in turn warms the air mass over the water. This cycle of warming and rising and cooling and falling is the relationship that provides the emergent behavior of high winds at sunset on the beach. The change on one side of the boundary results in a change on the other side—a transforming feedback loop.

Biological systems provide other rich examples of feedback. A classic example is the rise and fall of predator and prey populations in a food chain. Consider the rabbit and the wolf. As the number of rabbits increases, the wolves have ample food so their numbers increase, too. As the wolf population reaches a critical point, they consume sufficient numbers of the rabbits to inhibit growth of the rabbit population. Fewer rabbits means less food for wolves, so the wolf population decreases again. At one time, ecologists thought this system would result in an equilibrium—a balance—between the number of rabbits and wolves. The populations would stay constant at this equilibrium number until some other factor influenced the ecosystem. Using nonlinear computer models, scientists have realized even this simple ecosystem will not reach a stable equilibrium point. The populations will vary continuously as growth in one species leads to a growth or decline in the other. A change on one side of the boundary results in a change on the other side—a transforming feedback loop.

Breaking the Newtonian Model of Change

Notice that changes occur in both of these physical examples, but there is no single controlling factor either internal or external to the system. The interdependencies of the parts "control" the behavior of the system as a whole. The same is true of feedback loops in organizational settings: Each part of the system responds to changes in other parts, but no single factor determines the behavior of the system as a whole.

This mechanism for change in an organization is distinctly different from the one arising from Newtonian mechanics. In a Newtonian world, we expect one agent—manager, customer or stockholder—to define the change required for an organization. The agent exerts pressure on the rest of the system to enforce change. Much of our language about change in organizations is borrowed from the world of Newtonian physics: pressure, force, momentum, inertia and resistance. Our strategies, too, derive from an expectation of a single forceful agent for change: Find a champion, convince senior management, enforce expectations and anticipate and overcome resistance. Companies that depend on Newtonian models for change look for a powerful force for change internally, or they reach outside themselves to find consultants who are seen to "carry more weight" because they are external to the current structures. All of these efforts search for and depend on a single change agent or a compact group of change agents to define and execute change, as if the organization were a solid object to be moved from point A to point B.

In a complex, highly dependent system such as an organization, the expectation for a single-point cause of change is doomed to failure. Any individual change agent, regardless of its power or place in the system, is only one source of feedback among myriad competing feedback loops. Every part of the system—individual, team, department, corporation or industry—responds to feedback across an infinite number of boundaries.

Many of the messages received encourage stability and resist the change when it is pushed by the single change agent. No one participant in the system can override the multitude of messages that emerge from the accumulation of all other parts.

The parts of the system are linked by a multitude of individual and unique feedback loops. The complicated network of feedback loops is the closest thing to a control mechanism that exists in a complex system because the network carries messages and influence for change from one part of the system to others. Unless agents for change are able to transform the network, they will never be able to implement real change in the system's dynamics and behavior.

Regulating a Public Service

The mission of a state public service agency was to provide support to persons with disabilities. Services and products included education, counseling, job placement, mechanical devices and electronic tools. Historically, the agency had been divided into geographic regions, each of which was managed by a director. The directors were free, within budgetary constraints, to provide whatever products or services their counselors recommended. Counselors were free to work with individual clients, assess their situations and recommend whatever products or services deemed necessary. All of the counselors were trained professionals, but they received no centralized guidance regarding appropriate or consistent services for clients. Each made his or her decision based on interviews with a client and assessment of the client's file.

This system continued for years. The director in each region received only budgetary feedback from the central office, and counselors reporting to each director received feedback on monies from the director and feedback on needs from the client.

Regional directors functioned independently, so there was no feedback loop from one geographical area to another. Within this context, counselors granted or denied products and services to individuals.

As you can imagine, services varied widely across the state. Disabled persons with similar needs might receive completely different services depending on the region of the state they lived in and the counselor assigned to them within the region.

Because of this inconsistency of services, a new and powerful feedback loop was established through litigation. Consumer groups formed and sued the state, charging unequal distribution of resources. They also lobbied Congress to establish service-related criteria to regulate the activities of the agency. Introduction of these two feedback loops began a process of evolutionary change.

The agency formed a task force of selected directors, consumers, representatives of the central office and counselors. This group defined guidelines for assessment and allocation of services to clients. This process established strong transforming feedback loops among the team members, but excluded the rest of the agency's employees. When the team completed its task, the next step was to extend the transformation beyond the bounds of the working group.

To accomplish this, all of the regional directors were brought together for a series of working sessions. Over a period of months the directors came to recognize the need for consistency across the state. They discussed the new regulations, defined agency strategy and suggested tactics for implementing the new regulations. In short, they established transforming feedback loops among the regional offices, all of which previously had worked autonomously.

During this transformational process, all directors were charged with "carrying the message home." They were expected to transform not only themselves, but also their employees.

Through ongoing team meetings at the regional level and consistent case review and performance appraisal procedures, the directors were expected to transform all members of their teams. Using these new policies and procedures, they established feedback loops within the regions, where none had existed before.

The agency's central office contributed additional new feedback loops. It designed a two-day training conference for all agency employees. The conference provided consistent training and allowed counselors from different parts of the state to confer. It also provided a source of consistent reinforcement and an opportunity to celebrate the changes taking place in the agency.

The central office also instituted continuing training programs to help disseminate information, to provide evaluative feedback to all staff members and to collect information about how the changes were progressing. The office also established a consistent measurement system to collect, process and distribute information between the central and regional offices and among all regional offices. Each of these new feedback loops reinforced and consolidated the change effort.

Two other sets of feedback loops were established during this transformation. First, the regions recognized the need for the public to understand and appreciate the work they did, so each region created a public information function designed to meet the specific needs of their regional constituents. Second, they created brochures and orientation sessions to help potential clients understand agency regulations and client responsibilities.

This complex combination of feedback loops transformed the agency. No one individual or group caused the change. Everyone involved with the agency in any way participated. As members were transformed themselves, they transformed others. No one wrote a detailed plan for change. No one set time lines or goals for long-term success. No one stood on a soap-box to extoll the glory of change. Rather, the leader of the agency recognized the need to create new pathways for transforming feedback. She transformed the system's feedback-loop network, and the result was change. This is the only way substantial, permanent change can occur in a complex organizational system.

The Rules of Transforming Feedback

Certain criteria must be met before a pattern of communication can be established that will bring about the transformation that engenders productive change in systemic behavior. Not all communication qualifies as feedback. Not all feedback loops are transforming. Not all feedback generates the same results. The appropriate timing of feedback is critical to its success. Feedback criteria are described in the following sections.

Communication as feedback. Have you ever received a memo that seemed to come from left field? Usually it arrives late on a Friday afternoon, over the signature of some lofty executive. What does it mean? Why was it sent? What am I supposed to do with it? These are normal reactions to communication that is not feedback. Feedback must be clearly related to the history of a system or the parts of the system. It must be explicitly related to the previous performance of the system. That is why it is called feedBACK—the system's change feeds back into and has an effect on the system at a later time. If a message is received without context, then it is not sensitive to history and cannot qualify as feedback. It might shock the system, but it will not transform it. To be meaningful, feedback must come in, or at least with, a context.

Frequently we see this problem when Total Quality Management efforts are initiated in an organization. Senior management makes the statement, "From now on, quality is our top priority." People in the trenches respond, "For me, it has always been the top priority. Where have you been?" Such insensitivity to the real world serves to separate employees from management and from the Total Quality effort. If the message is not context sensitive, it will be meaningless, at best, or destructive, at worst.

Feedback for transformation. To be transforming, feedback must be bilateral. Both the sender and the receiver must participate actively in the transformation process. Before you can change another, you must be willing to change yourself. Only authentic, two-way communication will result in true transformation in the parts of the system and in the system as a whole. Many Newtonian managers miss this point, so they fail to "walk the talk." Their unilateral commands for change not only fail to produce the expected results, they foster resentment among employees. This realization is often a difficult lesson for a manager who has been successful using Newtonian techniques, but it is a critical lesson to learn.

Abe is the plant manager for a public utility. For four years he had tried to institute real change in his organization, but with limited results. After a particularly frustrating staff meeting, he went home disillusioned and distressed. Over a martini, he shared his concerns with his wife, who responded gently. "Abe, maybe you are the one who needs to change," she said warmly.

Abe says he's not sure whether it was a light bulb or a Roman candle that went off in his head, but at that moment he saw she was right. He had to change. The next day, he called his staff together and announced his intentions to change. He invited each of them to prepare a list of the ways in which he should change. Every day for a week, he met with one of his staff and discussed the list of proposed changes. Every evening Abe went home frightened and discouraged. Had he really been making that many mistakes? Could the plant continue to function if he broke his old patterns? Finally, and most important, would he be able to follow through on his commitment and make the changes his staff suggested? It was a dark and difficult time, but when the interviews were over, Abe listed for his employees the changes he intended to make and devised a process that would provide feedback on his progress. Over the next month, as Abe dealt with his personal transformations, he began to notice significant change in the plant's processes and functioning. The changes he had talked about and worked on for years suddenly happened. By the end of the year, all measurable criteria showed significant improvement. Abe had established authentic transforming feedback loops, and everyone else had responded in kind.

Feedback for results. The field of systems dynamics defines two kinds of feedback: Positive and negative. Each serves a different function, and both are necessary for the effective performance of a real-life organizational system.

Positive feedback gives the message, "Do more of whatever you did before." It amplifies the existing behavior of the system and tends to exaggerate outcomes. Any small change in the system that receives positive feedback will, during later cycles, be expressed more strongly. A squawking microphone is a good example of a positive feedback loop. A sound is amplified by the speaker, collected by the microphone and amplified again, until the sound is unpleasantly distorted.

You might hear examples of positive feedback in an organizational meeting when participants say

- Tell us some more.
- That's a great idea.
- I really like that.
- Good work!
- We can really use that approach.

Negative feedback gives the message, "Stop doing whatever you did before." It sets limits and takes action to ensure the behavior of the system stays within those limits. If negative feedback is given to a small change in the system, the change is squelched, and the system can return to a previous state. A room thermostat is a classic example of a negative feedback loop. If the temperature of the room moves outside an acceptable range, the system takes action to adjust. If it is summer and the temperature moves too high, the air conditioner turns on. If it is winter and the temperature moves too low, the heater switches on. In both cases, the natural trend of the system is reversed by the introduction of negative feedback.

You see negative feedback in meetings when participants say

- Let's get back to the agenda, shall we?
- Where did that idea come from (smirk, smirk)?
- That is too dangerous for us to try.
- We tried that before, and it didn't work.
- I don't think management would approve.

Both negative and positive feedback can be transforming. To function effectively, every organization needs both negative and positive feedback—some behaviors should be amplified and others should be squelched. The trick is to be selective, to use the appropriate feedback for the specific context, task and time. Overuse of positive feedback can result in loss of focus or attention to work. Overuse of negative feedback can result in despondency, anger or frustration. Balancing the two allows the organization to transform in constructive and adaptive ways.

Length of feedback loops. The time that transpires between initial action and resulting feedback is critical. If the loop is too long, then the context weakens and the feedback can lose its meaning for the system as a whole. A loop that is too short might not provide sufficient opportunity for transformation to occur before the responding message is sent. To be effective, feedback must be timely.

For example, a manager walks down the hall and notices an employee is playing a computer game during work hours. Negative feedback is clearly required, but the manager has several options for timing the feedback. He or she could make a note and say nothing until the employee's annual performance appraisal. Clearly, this feedback loop is too long. Both the manager and the employee will have forgotten the details of the event, and neither will be able to take action on the information at such a late date.

On the other hand, the manager might storm into the employee's office and demand he or she get back to work immediately. This is a short feedback loop, and in some circumstances it might be appropriate. In others, however, this precipitous action might be destructive. Given more time, the manager would have learned the employee had worked late the night before and came in early in the morning to meet a deadline. After working through lunch, the employee took a 10-minute break for a round of electronic poker to rest and recover before returning to the afternoon's tasks. The strong and immediate response from the manager would be viewed by the employee as unreasonable, unappreciative and controlling.

To be most effective, feedback loops should be long enough to ensure sufficient understanding and short enough to be associated with the initial behavior that resulted in the feedback.

Because any complex system includes many different parts and multiple levels of interaction, it will incorporate an uncountable number of feedback loops. The strengths and efficiencies of these individual feedback loops will always be determined by

- Their fit with the context of the system
- Their ability to transform both the sender and the receiver
- Their tendency to increase or decrease specific behaviors
- The amount of time required to complete the loops

You can think about feedback loops one at at time, and the previous analysis simplified it even further by focusing on only one aspect of a single feedback loop. In reality, however, any complex system includes an infinite number of loops, each of which fits into a unique context, relationship, effect, and time. The variety of feedback loops among its parts determine the strength and efficiency of a system as a whole.

The contents of the loops in a system must be varied because each individual loop must match its immediate context, and one system includes many different local contexts. If, for example, money is the only means of communication in a system, its feedback loops will not have the required diversity to support smooth functioning of the system over time.

All of the parts of the system that send or receive messages must be continually transforming according to feedback available to them. Such a mutual transformation allows for the flow of information and change to permeate the system and to unite the parts into a functional whole. For example, if management consistently refuses to listen and respond to messages coming from employees, then the systemic functioning of the system will be weakened.

Some behaviors that arise in the system should be amplified and others should be damped by a combination of positive and negative feedback. Managers must simultaneously encourage some behaviors and discourage others to maintain an adaptive and productive environment.

The cycle times for feedback loops should be varied, according to the contents and contexts of their messages. A combination of short-, medium-, and long-term feedback loops provide cohesion and coherence to an otherwise unstable system. For example, a manager might give some messages in immediate one-on-one interactions. Others are appropriate in regular weekly meetings. Still others should be saved for quarterly or annual gatherings.

By engineering variety into feedback, the manager is able to strengthen the bonds that hold the organization together. An infinite variety of transforming feedback loops provides the glue to hold the complex, dynamic system together in the midst of continual change.

Electronic Communications and Feedback Loops

Electronic communication tools have seriously affected the quantity and quality of feedback loops in today's organizations. Regardless of how they are used, the plethora of electronic communications might decrease the overall amount of face-to-face communication in which two-way, context-based feedback thrives. Over-dependence on electronic media restricts the pathways available for feedback, lessening opportunities for transforming feedback. Human contact remains the richest context for transforming communications.

Each of these tools has a place in the communication networks of complex systems, but the medium for a specific message must be carefully chosen to optimize the potential for constructive transformation. Only when an organization is conscious of its need to participate in transforming feedback loops will these powerful communication tools meet their potential as facilitators of transformation.

Consider the implications of communication mechanisms shown in the following table.

Mechanism	Benefits	Drawbacks
Voice mail	• Immediate satisfaction when leaving the message • Ability to broadcast • No waiting for someone to be available	• One-way communication • No reliable response
Electronic mail	• Two-way communication • Possible very short loop • Self-documenting • Good for concrete or fact-based conversations	• Short-hand language may distort message • Not good for abstract conversations • Depends on receiver to have and use compatible system
Desk-top publishing	• Generates professional-looking documents • Encourages long, one-to-many feedbacks	• May focus on form, not substance of message • May create long loops by delaying distribution
Fax	• Immediate satisfaction when receiving a fax • Relatively low cost • Provides hard copy information	• May not allow two-way communication • Generates pressure to respond immediately
Video conferencing	• Provides visual context for communication • Gives almost personal contact across geographical boundaries	• Relatively infrequent • Requires participants to go to a studio

What Can I Do about Transforming Feedback?

Recognizing the power of transforming feedback loops is the first step toward managing them wisely. After you have become aware of their varieties and their power, you will begin to focus on the loops in which you participate. This focus is the second step to effective management of feedback loops. Given such a focus, you can begin to take specific actions to leverage your communications into truly transforming feedback loops.

Tips on Transforming Feedback Loops

Every day you have numerous opportunities to establish or enrich transforming feedback loops in your organization. The suggestions below will provide you some guidance as you learn to take advantage of these opportunities.

Be willing to be transformed. Don't ever expect someone else to change at your bidding unless you are willing to be changed as well. This does not mean the distinction between the two of you will be erased, but it does mean you will make a good-faith effort to understand, recognize and adapt to the perspectives of others. Listen carefully to what others say and be conscious of your willingness to respond sincerely by adjusting your perspective to incorporate their perspectives. This, in fact, is the fundamental nature of what is currently referred to as the "learning organization." Because individuals are willing and able to learn and adapt, the organization learns and adapts over time. The process begins with the willingness of each participant, especially managers, to collect and respond to feedback from many different parts of the system.

Balance positive and negative feedback. The Newtonian models of management depend heavily on the use of negative feedback to maintain control. The complex-system manager, on the other hand, recognizes the success of the organization depends on amplifying individual's gifts and insights through positive feedback. Develop in yourself an ability to discern behaviors that should be reinforced and provide appropriate feedback for those behaviors.

Overcome your fear of the unexpected and resist the immediate urge to shut others down. Take time to investigate the context-laden value of their behaviors and to allow those surprising events to transform you and your group.

Pace yourself. Not all important messages must be delivered immediately. Delay or defer response to messages that might require extensive thought or creative problem solving. Let others know you are consciously putting off a response and why you have chosen that path. They will appreciate your candor and recognize you are taking time to consider an appropriate response. This tactic also will lend weight to your response when, at the right time, you send a well-planned message.

Don't expect to control all transformations. You participate personally in only a small fraction of the total number of transforming feedback loops of your group. Acknowledge that fact and learn to live happily with it. The only transformations you control are the ones that affect you. Do not think you can engineer the transformations of others. All you can do is to observe how others respond, adjust your subsequent messages to encourage transformations you desire and, again, observe the responses. This iterative process is one of adaptation, but it does not constitute control over the behavior of individuals or the system as a whole.

Look for variety. Never expect a single feedback loop to result in the transformation you seek. Use a variety of contexts, lengths and types of feedback to relay your messages. Adapt and adjust them over time to increase the likelihood of transformation. Be consistent in the message, but inconsistent in the medium. Send a single transforming message in personal conversations, group interactions, written and electronic media. Use regularly scheduled and random timing to deliver the message. Always adapt and adjust your delivery to take advantage of the rich environment of feedback loops that keeps your system dynamic and complex.

Building Feedback Loops: An Example

Mark's Market is an up-scale grocery store in an exclusive suburb of a metropolis. For years, the market's only competitors were the ethnic, old-town groceries in the heart of the city. Its customers had no desire to travel through traffic, noise and dirt to trade in the inner city, so Mark's developed a loyal following.

Economic and social changes in the early 1990s had devastating effects. Two-earner households resulted in less gourmet cooking. Traffic congestion made parking a hassle for customers. High-priced, fast-food restaurants sprang up in the neighborhood to lure customers away from Mark's. Prices went up and margins went down for several successive quarters, and the owner of Mark's was nervous. He decided he needed to communicate with his customers to find out what they wanted in a grocery store.

He considered placing customer satisfaction surveys on the check-out counters but decided such an intermittent and one-way communication would not give him the information he needed to transform. Understanding the mechanics of transforming feedback loops, he instituted the following measures.

Every day he obtained the name and telephone number of one customer. In the evening, he called that customer and chatted with him or her about why the customer had come to Mark's, what was expected of the store and whether those expectations were being met.

He spent one day each week wandering in the store's aisles. He noted traffic patterns and watched the faces of his customers as they squeezed the oranges and read the labels of canned goods.

He read everything he could find about the social and cultural changes he observed.

He ran a contest among his employees. Each was to create a display of goods that he or she thought would appeal to customers. He measured the effects of each display and gave a prize to the most successful.

Every month he tried a new experiment, advertised the change in the neighborhood newspaper and measured the results. He built a little play area for children while their parents shopped. He offered to deliver groceries to people's homes for a week. He covered the front wall with butcher paper and invited patrons to decorate the store with their favorite recipes. He gave cooking classes featuring quick, gourmet dishes.

This process allowed him to collect information he never could have gotten through a simple customer-service survey. The variety of tactics allowed him to establish and learn from myriad transforming feedback loops. Ultimately, based on the information he had gleaned, he redesigned his store and retrained his staff to meet the needs of his real customer base.

In this chapter, you have learned about the special kind of connections that bind a complex system into a functioning whole. You also have discovered how you can engineer and participate in transforming feedback loops to introduce permanent change into your organization.

Chapter 4
Fractals

Chapter 4
Fractals

Nature has a wonderful way of creating complex, adaptive systems. It begins with a small piece of matter that contains a code for the whole organism. Then, it follows the code to draw in and transform material from the environment to create a whole new entity. This transformation process is iterative—the first step is repeated many, many times, with slight variations for each cycle. The first cell divides into two, each of them divides into two and on and on. Sometimes the new cell is identical to its parent and sometimes it is different. Scientists understand only some of the mechanics of this process, but we all know it works. A seed, planted and watered, germinates and moves toward the sun until it looks, for all the world, like its parent plant. A zygote divides and replicates to create a fetus, and the fetus grows until it emerges as a mature being.

This natural process of "coming to be" begins with a simple seed and, through complex replications, generates an object of infinite complexity, functionality and beauty.

Chaos mathematicians use this same process to generate a kind of geometrical object called a fractal. Chaos managers also can use this process to generate human organizations that are complex, functional and beautiful.

This chapter describes the process used to generate a fractal and demonstrates how organizational structures can be developed in the same way. It also describes the complex nature of the resulting fractal and describes how you can use these fractal characteristics to generate effective organizational structures.

What Is a Fractal?

Human life begins with a complete set of DNA, and the fractal begins with a simple, nonlinear equation. These components constitute the code or pattern from which the organism or the fractal will be generated through a series of replications and variations. Biological systems complete this process according to a set of rules, only some of which are currently understood. Mathematicians use supercomputers to select individual values for one of the variables in the nonlinear equation, solve a series of iterations using that value, test the stability of the solutions and plot a colored dot on a plane to represent the stability for that value when used in the equation. Then, they select another value, solve repeatedly, test and plot the outcome. This process is repeated until all of the numbers represented on the plane have been tested and plotted.

This summary of how a fractal is generated is sufficient for us to work with the idea as a metaphor for organizational development. If you would like more information about how a fractal is generated, refer to *The Turbulent Mirror* by Briggs and Peat.

Why Generate an Organization as a Fractal?

What would it mean to generate an organization in the same way a fractal or a living thing is generated? To create a fractal organization, the leader begins with a specific and clear concept of the organizational principle. The principle might be personal achievement, customer focus, technical superiority, interpersonal relationships, competition, profitability or any fundamental concept of business or human interaction.

This basic principle is applied in every circumstance the leader encounters. A well chosen principle will not always result in identical actions, because the iterative application is always sensitive to the immediate environment in which the action is taken or the decision is made. The principle is applied to immediate circumstances and information the leader has at hand. It is not rigidly applied to enforce the same result, regardless of the environment or immediate situation.

Employees are taught, either by modeling or explicit instruction, to apply this same principle in all things. Customers come to expect all of the organization's behaviors to be generated from this same, relatively simple, source. Vendors, too, respond to the underlying principle as they see it affect their interactions with the company.

The common idea lends coherence to the organization, because it provides an underlying similarity to all corporate actions. It provides for some form of predictability, even in the face of constantly changing circumstances. It sets a reliable context in which all corporate and business decisions can be made.

An example will help show how this process works. Excel Instruction, Inc. is a technical training and documentation company started by my business partner and me in 1986. Training is our business, and our underlying principle of operation is learning. Our fundamental value and reason for being is to learn and to teach. Learn and teach. What does it mean for our organization to adopt this as the "seed" for our fractal? Every question that arises is approached in terms of what we or others can learn in our business interactions.

Business Activity	Learning Fractal
Sales calls	Multi-seminars in which our clients teach us about their businesses and their problems and we teach them about our products and services
Customer service	Learning to meet customers' needs more effectively
Performance appraisals	Opportunities for us to learn about the perceptions and aspirations of our employees and for them to learn about our evaluation of past performance and expectations for future activities
Quality control	Information feedback to employees while learning more about the specific requirements of customers
Staff meetings	Chance to share discoveries with others
Business plan	Evolving document that incorporates our discoveries about the market, our resources and product possibilities
Teamwork	Encourages cross-training and individual development on a project-by-project basis
Project and financial reporting	Allows individuals, teams and the organization as a whole to learn ways to become more efficient and effective

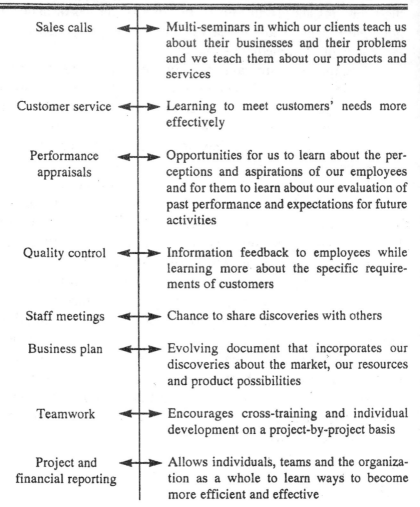

All of our internal and external conversations and all of our policies and procedures are established to optimize learning by individuals and by the organization as a whole. Every iteration is unique because it responds to the immediate situation, but all iterations have a fundamental similarity because of the corporate commitment to learning and teaching.

Regardless of the specific chosen principle, an organization that has a clear generative principle and that applies the principle consistently will demonstrate both coherence and adaptability. Coherence comes from the basic pattern, and adaptability will come from the context sensitivity of each iteration or application of the principle.

Now, how does this complex, chaotic process differ from the traditional, Newtonian one? In the old models, a new organization was seen to be designed like a building. The initiators drew a blueprint, which included all of the components of the building and defined how the parts related to each other and the environment. This blueprint defined the stable limits of the parts and the whole, and it became the definitive pattern on which the whole was built. This process worked fine when the components of the organization and its environments were predictable and simple. When the organizational structure supported a limited number of functions, personnel were relatively homogeneous, economic environments were stable, and customer expectations were consistent from year to year, the blueprint model worked exceedingly well. Leaders could engineer a corporate solution to survive the minimal internal and external variations expected in the near- and long-term future.

When such an organization encountered changes, however, major problems arose. If the employees were no longer willing to fulfill their restricted roles, if the market changed dramatically, if technology provided new ways to perform old tasks, then the whole corporate structure had to be destroyed. The company would begin with a whole new blueprint and expect the existing structures to rearrange themselves in one cataclysmic reorganization. This approach was not only wasteful, it frequently was impossible. Within a short time, the old organizational structures returned, like old stains bleeding through new wallpaper.

The complex, fractal model, on the other hand, provides a flexible structure that is not defined by its boundaries but is determined by its core. The core stays constant in the midst of environmental change, but the specific structures and interactions of the parts within the organization and between the organization and its environment are in constant transformation.

What Is the Result?

The generative process is not the only special aspect of a fractal organization. The fractal process creates a resulting structure that has some exciting and functional characteristics. In this section, we will describe the characteristics of a mathematical fractal and explain how these same characteristics appear in organizations.

Self-similarity. When you look at a fractal, you will notice different parts of it look similar to each other. A swirl here looks like a swirl there. Patterns of dark and light repeat. Splashes of color appear to be of similar shapes in different parts of the structure. The repetitions might not be identical, in fact they frequently show subtle differences, but they are recognizable as parts of the same whole.

The same self-similarity appears in biological organisms. The most common comments in a hospital nursery refer to self-similarity, "She has her daddy's nose." "His grandmother had the same red hair." "Look at those dimples, just like her mother's." Because they share DNA, family members exhibit marked self-similarity. We are able to distinguish between an oak tree and an elm because the leaves of each carry the self-similarity of the type. We expect and find related organisms to share certain traits. Each individual member will incorporate the self-similarity within a different combination of characteristics, but the similarity is still detectable.

Self-similarity in an organization is no less apparent. When you walk into a new organization, you will notice similarities in behavior across the entire corporation. Different departments in the same corporation will share values, procedures or communication habits. Individual employees adopt each others' forms of dress and behavior. Teams will replicate the interpersonal norms of other teams. This self-similarity is explicit even in trivial situations. For example, you will seldom find an immaculate coffee area on one floor and a filthy one on another floor of the same company. The behavioral norms are self-similar across the organization.

In an organization that is consciously generated through iteration of a basic principle, self-similarities are constructive and consistent with the explicit expectations of its members.

A rigid dress code, for example, ensures that members of an organization look like they belong. Employees recognize themselves and others as parts of a functioning whole. Communication symbols, such as logos, allow customers and employees to discriminate between one company's products and those of its competitors. Company rituals and stories bind an organization together by defining what all of the individuals share, in spite of the differences in their skills, talents or organizational levels.

Scaling. The fractal has levels of complexity that are not immediately visible. To see them, you must magnify the image. If you take one small part of the image and magnify it, you will find that the new image shares self-similarity with the original one. Magnify the new image, and you will find that the image still exhibits self-similarity. Theoretically, you could continue to magnify the image and find an infinite number of self-similar structures. This similarity across levels of magnification is called "scaling."

Broccoli is an elegant example of scaling in nature. A whole head of the vegetable has a characteristic tree-like shape. One branch reflects the same structure, and a branch of the branch also reflects the whole. This repetition of shape across scales gives the plant a coherence in spite of its complexity.

You find evidence of scaling in organizations when the ethical behavior of an individual employee reflects that of the leader, when corporate values are evident in individuals' behavior, when media coverage of the organization influences the way employees feel and think about their place of work. These scaled factors link the parts of the organization into functioning wholes.

When critical characteristics are not scaled in the organization, for example when a leader behaves in one way and expects different behavior from employees, dangerous rifts develop in the *esprit de corps*. Groups develop distrust and lose a sense of loyalty to an organization that disregards the need for scaled values and expectations.

Creative differences. Self-similarity and scaling are not the only things giving the fractal its beauty. Within the context of repetition, the fractal introduces stark differences. These contrasts set a framework in which the similarity is apparent. Difference in the midst of similarity and similarity in the midst of difference create what we can perceive as an artistic whole.

Natural beauty also emerges from distinctions between the same and the different. A face is beautiful because of its symmetry and its surprises. A mountainside of grassy meadows, evergreens and deciduous trees provides a panorama that can be breathtaking. Dancing flames in a fireplace mix patches of gold and blue and red into alluring patterns.

These beautiful differences are functional as well. Symmetry allows for efficient growth and change. Mixed vegetation supports a stable ecosystem. Different gases burn in different colors and at different rates to ensure the fuel is used efficiently.

Difference in the midst of self-similarity provides efficiency and flexibility as well as aesthetic appeal.

Organizations, too, thrive on difference. Specialization of function allows individuals to perform tasks at which they excel. Cultural differences introduce new perspectives and creative insights. Separate offices give employees an opportunity to express their individual tastes and introduce personal and family symbols into the workplace. Multiple hierarchical levels give explicit attention to the need to manage the whole as well as the parts of the organization.

A successful merger of two companies depends on the ability to recognize and manage differences. Distinctions between the two institutions cannot and will not be erased by fiat. They must be integrated into a complex picture of a whole. A well-managed merger will include the processes of identifying the different strengths each of the partners brings to the transaction and consciously designing ways in which those strengths can be integrated into a diversified whole.

Fuzzy boundaries. As you look at a fractal, you will note there are no clear lines of distinction between one part of the image and another. Instead, each line of demarkation is jagged and indistinct. In fact, if you magnify the boundary, it becomes even more complex and fuzzy.

This same phenomenon appears in nature. We talk about genes as they appear on chromosomes in the DNA molecule, but there is no physical boundary between one gene and the next. Different parts of the molecule function differently, but there is no way to tell where one gene starts and another ends. We might think of cell walls as distinct separations between inside and outside, but, when magnified, the cell wall does not form a clear line of distinction. Instead, it is a complex combination of molecules that allows for materials to be absorbed and expelled.

Organizational boundaries are fuzzy, too. The formal organization chart might make it seem possible to distinguish between one functional group and another, but if you watch real, day-to-day interactions, you find that lines of communication and loyalty are fractal in nature. Few organizational boundaries are simple straight lines dividing one group from another.

The nature of human communication builds fuzzy boundaries in spite of management's efforts to clarify distinctions. In a Newtonian, blue-print-based organization, these fuzzy boundaries are seen as a problem to be controlled. In a fractal organization, they are seen as the focal points for creative interaction and transformation.

Good customer service, for example, requires a fuzzy distinction between customer and vendor. The vendor must empathize with the customer enough to understand his or her concerns, while the customer must be cognizant of the limitations of the vendor's products and services. Honesty on both sides breaks down a rigid boundary and generates a fractal one in its place.

Variable stability. Some parts of a fractal are more stable than others. Different parts of a fractal represent different levels of stability. Think of the leafy fern as an example. The stems are more solid than leaves, and larger leaves more stable than small ones.

The different patterns in the image represent the stability of the initial equation. If you can move a long way from an initial point and find similarity, then the initial point is stable—a lot of change results in little difference in outcome. If you move just a small bit and find a very different patch, then the initial point is unstable—even a small change will result in an extremely different outcome.

Nature exhibits both areas of relative stability and instability. One example is the weather. Some weather systems, such as a powerful high pressure area, change little, even over a wide area. In other places and at other times, a short walk will move you into a different weather condition. Consider the erratic movement of a tornado, in which one house might be destroyed, while the house across the street suffers little or no damage.

Another example is the motion of a golf ball on a green. Depending on the contour of the green, a ball might land in one place and roll from the hole, or land a short distance away, and either stop entirely or roll toward the hole. The behavior of the ball depends on the configuration of its immediate environment.

We find this variable stability in organizations, as well. A slight change might have no effect in one part of the organization, but generate significant change in another. In general, areas of the system closest to boundaries are most likely to generate large changes.

For example, consider a small change in behavior, such as saying "please" and "thank you" consistently. If this behavior is adopted in the accounting department, it might make no apparent change in the financial behavior of the company. The same change among those who provide customer service, however, might generate increased client satisfaction and market share. Similarly, a new computer system might generate significant improvement in accounting performance but have little direct effect on customers.

By thinking of your organization as being generated as a fractal is, and if you learn to recognize and use the fractal characteristics of your group, you can anticipate the outcomes of actions and adapt more effectively to changes as they occur.

Are All Organizations Fractal?

Yes, all organizations are fractal. They all take a small set of operating principles and apply them in numerous, unique situations to generate an overall pattern of behavior. The question is:

Does management recognize and take advantage of this natural propensity?

Management can ignore the generative behavior of the fractal organization and the characteristics it creates. Newtonian managers expect the organization to develop from the blueprint they created. They can overlook or overuse self-similarity, ignore or destroy the fuzzy boundaries that exist, encourage destructive differences and disallow constructive ones, and assume the same amount and kind of change will have the same effect across the organization. Historically, this ignorance of the fractal nature of organizations has not seriously affected institutional health or well being.

But in tomorrow's fast-paced world, ignorance of fractal growth and characteristics will stymie management. Leaders will be left without an inkling of how their organizations evolve and how to harness that evolution to improve performance and profitability. The next section describes specific ways you can work with the fractal nature of your organization to facilitate adaptation and change.

What Can I Do about Fractals?

The fractal model provides a new and different way to think about your organization and how its parts fit and work together. If you want to understand how work really gets done and how meaningful, permanent change can happen, you must ignore the formal organizational charts and see the fractal structure of individuals and groups within the company. The following strategies will help you make this shift in understanding and develop behaviors to take advantage of the organizational fractal.

Fractal Strategies

Your organization has evolved using a fractal process—multiple applications and repetitions (iterations) of a fundamental principle. It also exhibits many of the characteristics of a fractal structure. To be an effective manager, you must recognize and use the forces and features of fractals. The suggestions below will help you meet this challenge.

Recognize change through evolution. The thought of a quick and total change for an organization is enticing. Wouldn't it be nice if we could imagine a new future and have it immediately? In a complex, chaotic system, such a change simply is not possible. If change comes, it comes through individual adaptations to individual messages received over time. To be an effective agent of change, you must realize that.

Your organization has evolved into its current state and will continue to evolve, whether or not you intervene. The best you can do is understand the current state as a fractal image and begin to introduce new or different generative seeds or manipulate the iterative applications of principles in specific parts of the organization.

Once you have recognized this complex, chaotic reality, you will be free to participate in evolutionary change, much as the farmer participates in the ripening of a crop.

Find the seed. What is the fundamental principle that has generated your current structure? Has one principle affected all parts of the organization, or does each division have a different fractal starting point? Some of the traditional ways of stating values and corporate mission and vision can help you identify the seed. But be careful. A mission as stated by senior executives might have no relationship to the reality of the organization. You must listen to employees, watch their interactions and identify the ways in which they agree and disagree before you can see what the real starting point is for your fractal generation. Be sure the mission and vision reflect the current reality, not what a small group thinks the mission or vision should be.

Use the seed to generate change. After you have identified the basic principle on which your organization acts, you can begin to use the same principle to encourage change. This process can take many forms and should be focused on specific situations and questions. The following tactics—based on the fractal seed—will help you replicate change faster.

Make the seed explicit. Discuss the principle in meetings, put it on posters, share it with customers and vendors. Talking about the central value of the company helps employees see it and respond with their own constructive iterations. As they use the principle to make decisions, they will see its power.

Collect and tell stories about the organization. These might be corporate myths about the beginning days of the company or simple success stories about a project or individual. When these stories are based on the operating principle, they help everyone understand the infrastructure on which the organization is based.

Incorporate the principle into problem-solving procedures. If you have a corporate procedure for conflict resolution or group problem solving, insert a specific question based on the fractal seed. For example, at Excel, we always ask the question, "Who will learn and what will they learn if we adopt this solution?" Again, this explicit use of the fractal seed will help everyone learn to think within the complex structure of the organization.

Integrate the seed into all parts of the organization. The seed will affect all parts of the organization, so think about how it will affect marketing, distribution, product development, customer service, sales, manufacturing and support services. Ask each group to consider how it uses or might use the principle to do its work.

Be consistent in your own work. Employees will respond more readily to what you do than to what you say. Be sure your decisions are consistent with the fundamental principle. Think about it. Talk about it. Use every opportunity to reinforce the applications and the power of the fractal seed.

Each of these tactics will help you identify and leverage the seed of your organizational fractal.

Recognize and reinforce variety. The principle will provide the self-similarity your organization requires to be coherent, but variation in application provides the adaptability. Encourage creativity and cultural differences within the context of the system, but, at the same time, be sure the principle remains the same. Refer each action and each decision back to the principle, but also allow each one to reflect its immediate circumstances and individual strengths of employees.

Acknowledge fractal boundaries. Think about each of the boundaries that affect your organization and determine to what extent each is a fractal boundary. Is it clear and distinct, or are there gaps through which information can flow? You should also focus on participating in fractal boundaries yourself. Establish relationships that transcend the lines of the organizational chart. Participate in cross-functional groups whenever possible.

Introduce change at unstable points of the system. After you have evaluated the current fractal characteristics of your system, you will have an idea about which departments, groups or individuals are more likely to respond to change. When you wish to introduce a change, do it first with them. A little effort applied near a fractal boundary can facilitate change that would be impossible in another segment, even with unlimited resources.

Fractal Organization: An Example

SComp is a major player in the field of supercomputers. The company consists of three, semi-independent divisions that are separated geographically. One handles research and development of new machines, one handles manufacturing and one provides software development and customer support. The company was founded in the early '70s by an engineer who was and is considered to be a genius in the design of high-speed computers. In the early years, the company was built around his pioneering designs. He was an inventor who surrounded himself with inventors, so the idea of individual technical discovery became the fractal seed around which the organization was built. Individual, technical discoveries were expected, encouraged and rewarded.

The company succeeded, winning a large market share and growing rapidly through the mid-1980s. Eventually, however, problems arose. The research and development division formally broke away from software and support. Stock prices rose only marginally. Two rounds of layoffs and multiple reorganizations shook customers' and employees' faith in the organization. The reasons were many and open to interpretation. Some contributing factors are listed below.

- Entrance of competitors into the supercomputer market
- Internal strife regarding optimal technical solutions
- Lack of efficient and effective customer service
- Absence of applications software to run on the machines
- Saturation of a market in which not every company needs a super-computer

Based on its fundamental principle, the company took what it thought was the proper action to counteract each of these issues. It hired people believed to be the most talented designers in the world. It supported development of alternative architectures in parallel, assuming such internal competition would ensure optimal solutions to technical issues. It staffed customer service and support areas with persons who were inventors in their own rights. It funded the development of complex, state-of-the-art applications, so the software products would be as innovative as the hardware products. It invested in small companies with the ability to create innovative software applications. And it introduced new lines of hardware targeted at smaller, less expensive applications. In short, it approached all of its problems with the assumption that individual discovery and invention was the only SComp solution.

What it did not consider were the negative implications of the fractal seed it had selected. During the late '80s and early '90s, problems continued to mount.

- Marketing activities were focused on the technology rather than the customer or application. Product information and sales procedures were helpful only for customers who were, themselves, technically sophisticated.
- Large-scale projects required cooperation and teamwork, and team rhetoric abounded. The problem was that employees had been selected on their ability to work as independent agents and inventors. Their personal and professional skills did not fit well into a teaming culture. Many projects became battlegrounds of egos and technical strategies.
- World-class experts in a subject domain, such as meteorology, were expected to work closely with world-class experts in software design. The result was distrust and distorted communication among team members.
- Every product was developed as a creative effort. Individuals would solve problems in innovative but idiosyncratic ways. As a result, maintenance and installation were difficult and inconsistent. The unbridled creativity also made it difficult to keep training programs and technical reference manuals up to date.

These and other effects flowed directly from SComp's selection of personal expertise as its fractal focus. Yes, innovation was a powerful seed for this company, but SComp failed to recognize or adjust to the weaknesses engendered by the seed. Management worked with a world view that intended to control the external boundaries of the organization and allow the internal boundaries to be controlled by individual competition. This paradigm kept managers from recognizing and influencing the real evolution of change in the company.

SComp has reorganized many times to try to enforce cooperation among the technical groups. These reorganizations will fail until the company finds a way to change its fundamental principle or to use the principle of innovative competition to better corporate benefit.

SComp applied its fractal seed in all aspects of work, even when it was not constructive to do so. Understanding the fractal nature of your organization will help you avoid this trap and will help you select effective and efficient management activities.

Chapter 5
Attractors

Chapter 5
Attractors

When does traffic frustrate you?

You are driving through beautiful country on a scenic state highway. Occasionally you come upon an oversized farm vehicle moving at forty miles per hour. You slow, too, and take more time to enjoy the sights. Not much frustration there. Farther down the road, however, you encounter a detour that routes you to the near-by interstate and separates you from the unique beauties of the landscape. The detour irritates and distresses you. In other circumstances, though, you might have selected the high-speed interstate to reach your destination.

You are late for an appointment, and the usual exit is under construction. You have to travel three miles out of your way and, as a result, you arrive at your destination an extra five minutes late.

Driving downtown, you grow accustomed to the regular pattern of stop lights. Stopping and starting according to this accustomed rhythm is not distressing, but stopping for an occasional pedestrian or emergency vehicle can cause your blood pressure to rise.

In all of these stress-inducing situations, you began with an expectation of a pattern of traffic behavior, but some unexpected occurrence disrupted the pattern. Your frustration arises from the disrupted pattern, not from the immediate source of the disruption.

The same is true of patterned behavior in organizations. If you expect one pattern, and the system behaves according to that pattern, then you experience low levels of stress. If you expect a pattern and see another, then you experience frustration. For this reason, it is imperative that you understand the patterns that shape the behavior in your environment and learn to anticipate the behaviors that the patterns create.

In complex systems, patterns of system behavior are shaped by attractors. Scientists have identified three types of attractors that shape how objects in the system respond to stimulation and movement over time. The part of the system that is influenced by an attractor is called the attractor's regime. This chapter explains the three types of attractors, demonstrates how the attractors can be seen in organizational settings and provides guidance to help you work effectively within the variety of attractor regimes.

What Is an Attractor?

A complex, chaotic system is made up of a large number of interdependent subsystems. Each subsystem is independent in some ways but is influenced by other subsystems through a series of transforming feedback loops. As a result of this complex interdependency, individual actions of the parts are molded together into the behavior of the whole system. You can observe and understand the behavior of one part of the system, but it might not give you information about the pattern of the whole. On the other hand, you might understand the patterned behavior of the whole system, and not be able to predict the behavior of an individual part at a given time or place. A woven tapestry demonstrates the difference between the piece and the whole pattern. The individual system component is like a single thread woven into the tapestry. You can observe its color and texture, but it does not tell you about the pattern of the whole. Alternatively, you can appreciate the pattern of the whole, only when you shift your focus away from the individual threads that form the pattern. Attractors are the chaos science tool for describing the behavior of a whole system. To understand and describe a system's behavior, you must observe the behaviors of both the individual and the patterned whole.

Consider the stock market, for example. You can watch the changes in the Dow Jones Industrial average to see how the market performs as a whole, but this will not tell you the changes in price for an individual stock. Or, you can watch a single stock closely, but you will get no reliable information about the overall condition of the market. The accumulation of individual stock performance influences the performance of the market, and the market influences the performance of an individual stock, so there is a clear relationship between the macroscopic and the microscopic. But, understanding one will not ensure an understanding of the other.

Much of management science provides techniques for understanding and influencing the behaviors on the microscopic level. Performance feedback, walking around, individual observation and quality control measures all provide detailed information about the performance of individuals within the system. These approaches do not, however, provide much information about the patterned behavior of a whole team, division or corporation.

Newtonian science assumes group behavior is simply a summation of the behaviors of the individual parts. If Joe does one thing and Sally does another, you should be able to add their actions together and get a reliable picture of the whole. If the system includes too many Joes and Sallys, then you might have to use statistical tools to sample individuals and to construct the additive picture at the macroscopic level. Essentially, though, you are adding up the parts to get a picture of the whole.

In a complex system you cannot just add the parts together and assume you have a reliable picture of the whole. The interdependencies of the parts gets in the way because the parts affect each other as they are combined. The whole might be either greater than or less than the sum of the parts, depending on how the parts influence each other. If Joe and Sally have a common history and complementary skills, together they might create a high performing team. On the other hand, if Joe and Sally are quite different and they have competing skills, their combination might lead to less effective performance than either one would provide alone.

A new set of tools is required to consider the patterned behavior of a whole in a complex system. Chaos scientists use attractors as a tool to describe the overall behavior of complex, interdependent systems. There are three general classes of attractors: Point, periodic, and strange.

Point Attractors

In a point attractor system, all activities in the system tend toward a single point. The classic example is demonstrated with a round-bottomed bowl and a marble. No matter where you place the marble along the rim of the bowl, it will eventually come to rest at the bottom of the bowl. The marble might roll around and follow unexpected paths, depending on where you start it and how hard you push it. But you can bet, at some point in the future, the marble will stop moving, and it will be located in the lowest point.

You also see point attractors in traffic patterns. The speed limit on a super highway acts as a point attractor. Drivers will increase speed until they reach the posted limit (or ten miles per hour above the limit). If a single driver exceeds the limit, he or she should slow down again to stay within a reasonable range of the point attractor. When you drive on a highway, you make the reasonable assumption that everyone else will abide by this attractor. Any deviation, such as a slow-moving vehicle or a real speed demon, disrupts the pattern and stresses the other individual participants in the system.

Point Attractors in Organizations

An isolated team with a clear goal can work toward a single point outcome. Small, single purpose work teams, such as skunk works, are examples of consciously-defined point attractors in a business environment.

Start-up companies led by strong entrepreneurs frequently behave as point attractor regimes. Later, when the start-up goal has been reached, the company loses the point attractor and wanders aimlessly unless another attractor regime is clearly defined.

Hierarchical organizations frequently function with advancement to the next level as a point attractor. In such situations, every behavior of every employee is focused on gaining the skills and visibility required to move up the ladder.

As the examples show, a point attractor in business can be productive and effective or distracting and disruptive to the organization as a whole. One of the roles of the good manager is to decide whether a point attractor is the right overall pattern of behavior for an organization at one place and time.

If you expect point attractor behavior from your system, but you observe something different, you will experience frustration and the organization as a whole will be stressed.

Periodic Attractors

In a system driven by a periodic attractor, behavior tends to repeat in regular time intervals. The classic example of a periodic attractor is the motion of celestial objects. On a regular time table, the moon moves around the earth. Periodically, it returns to the same point in the sky. It waxes and wanes in its monthly cycle. This periodic motion is variable because change occurs, but it is reliable because the same changes repeat regularly.

Downtown traffic, with its regularly timed stop lights, is a wonderful example of a periodic attractor. The traffic moves and stops and moves and stops. When you drive downtown, you expect the regularly repeating cycle. The stop-and-go motion that would distress you on the highway is an expected part of city driving. On a more macroscopic level, traffic patterns in a whole city exhibit periodic attractors when morning rush hour traffic moves toward the city center and evening rush hour traffic moves away. Cities that have a seasonal appeal to tourists experience marked periodic changes in their traffic patterns, as well.

Periodic Attractors in Organizations

Seasonal products and services enforce periodic attractors on all aspects of corporate structure.
In a democracy, government agencies respond to election cycles on a periodic basis.
Financial record keeping enforces periodic cycles onto any business through monthly, quarterly, and annual reports, billing cycles and payrolls.
From the longer view, personnel cycles are periodic. New employees learn the ropes, work in maturity and ultimately retire.

You will notice some of these periodic attractors are internally focused and deal with policies and procedures. Others are externally focused and depend on customer and product availability. A good manager must distinguish between the internally- and externally-driven cycles to ensure the balance is conducive to customers' needs and corporate performance.

Within the general category of periodic attractors, there exist slight variations called "quasi-periodic attractors." When a quasi-periodic attractor reigns, the time required for a cycle might vary or the behavior of the system does not return exactly to its previous state. Business examples include economic cycles, corporate reorganizations and product life cycles. Sometimes it is difficult to recognize or anticipate the behaviors of quasi-periodic systems. In fact, they frequently emerge as a system moves from a simple periodic pattern into the third type of attractor—the strange attractor.

Strange Attractors

In the domain of the strange attractor, the behavior of the system appears to be random. No clear single point or regular cycle is immediately evident, but if you consider the system-wide behavior over time, complex patterns are discernible. Chaos scientists have identified strange attractors in many different physical environments. One of the most interesting is the dripping faucet. When the sound keeps you awake at night, you might think, at first, it is periodic—drip. . . drip. . . drip. A moment later, it might appear to be random—dripdrip . . . . . drip. . .drip. As you listen, you search for the pattern, but the pattern eludes you. The pattern is unclear because it exists outside of time. Scientists have plotted the time sequence of a dripping faucet and found it exhibits an extremely strange pattern. The pattern has a recognizable, but not repeated, shape. The system never returns to its original state, and the behavior at any instant cannot be predicted, but a pattern does exist.

This patterned randomness appears only in complex, interdependent systems. Newtonian science, with its dependence on additive relationships between parts and wholes, has no way to account for such behavior. So, this attractor was named the "strange" attractor. Though it is outside the bounds of classical mechanics, the strange attractor is not rare. Many physical phenomena, such as the heartbeat, electrical brain activity, weather systems and turbulence in fluids, are currently understood to work within the regime of strange attractors.

In traffic patterns, you will recognize a strange attractor in the area of an accident on a major highway. Traffic on the opposite side of the road slows down while drivers gawk. Even hours after the road has been cleared of debris, the traffic patterns in the area are distorted by the strange attractor that began with the accident.

Few strange attractors have been experimentally verified in organizational settings. The number of variables in functioning human systems makes it difficult to identify and collect relevant data, and long time series are required to discern the patterns emerging from strange attractors. Given these complex measurement issues, most of the data indicates that behavior in organizational settings is random. But, intuition indicates corporate behavior is not random at all. Patterns in behavior emerge over time. Some of those patterns can be described as point and periodic attractors, but others don't fit either of the predictable molds.

Whether or not these patterns can be measured and verified as strange attractors, is not relevant for the practicing manager. What is relevant is that the behavior of the group follows discernible patterns, even if the patterns cannot be predicted or controlled. Working within the patterns is satisfying and effective, in the same way that driving slowly on a scenic route or driving fast on a super highway seems satisfying and effective.

Working against the pattern is frustrating, unproductive and dangerous, in the same way that paying more attention to the speed limit than to the prevailing speed of your fellow travelers courts danger. In the next section, we will look at some of the characteristics of strange attractors that make it possible to take meaningful action in an attractor regime.

Stranger than Fiction

In distinguishing between point, periodic and strange attractors, we focused on only a few of the characteristics of the strange attractor (apparently random behavior, underlying patterns, lack of repetition of behavior, and lack of regular, periodic states). These characteristics allow you to discriminate between other attractors and the strange attractor, but they don't give you enough information to distinguish between random, unstructured behavior and the strange attractor. In this section, we will investigate other, special features of strange attractor systems that will allow you to distinguish between randomness and strange attractors. Those characteristics are described below.

Multiplicity. The current understanding of strange attractors indicates they emerge when a system is under the sway of multiple point and periodic attractors and when random noise is introduced into the system. This might sound pretty complicated, but it does describe many experiences in the real world.

Consider, for example, a live orchestral performance. Point attractors are at work to bring all of the musicians together at the right time and place to perform the same piece. All of the professional musicians focus attention on the conductor. The audience arrived because the place and time were announced. In the midst of these point attractors, periodic attractors are also at work. All of the performers count measures in a uniform period-

Strange Attractors in Organizations

Authentic commitment to customer service forms a pattern in which individual employee behaviors cannot be predicted, but they consistently respond to individual customers' needs.
Organizational culture describes an emergent pattern affecting the behavior of all employees. An effective culture transcends the formal statements of values and mission to shape the actions of individual employees in their respective locations and tasks.
Systemic fear, gripping an organization in times of uncertainty, forms patterns of behavior similar to the eddies of turbulence at the bottom of a waterfall. The pattern includes areas of relative calm and others of violent motion.

ic cycle. The composition itself provides periodic patterns of repeated melodies, harmonies, and modulations. The audience responds to the program with expectations of closure within a movement and movements within a piece. These point and periodic attractors merge together into the musical experience. In addition to these patterned components, the evening includes random events that contribute to the overall richness of the experience. These might include friends met during intermission, a slight variation in pitch from one of the violinists that affects the pitch of the orchestra as a whole, coincidences of the day that allow the cellist to falter imperceptibly, a broken string, an individual in the audience who has a momentary distraction, changes in humidity affecting the behavior of the instruments. These and other random occurrences join with the predictable point and periodic attractors to allow underlying patterns to emerge and make this experience a unique one for each participant. The evening as a whole might be described by a strange attractor, while each of the component parts contributing to the experience represented point, periodic or random events.

The same kind of convolution of components forms strange attractors in business environments. On a single day, the performance of a team might be affected by point attractors (specific, clear objectives or scheduled meetings), periodic attractors (day of the week or month of the year or stage in the team methodology) and random events (personal experiences of team members or a story in the newspaper). All of these separate patterns merge to shape the behavior of the team at a particular place and time. The overall pattern might not be easily discernible, but it clearly transcends the summation of the component patterns.

The interdependencies of the various attractors can be highly complex. Any one of the components might be particularly helpful or disruptive at a given instant. The focus of the whole moves dynamically from one attractor to another over time. The outcome might appear to be random, but it is the result of a large number of interacting, disparate forces. The only way to describe the overall pattern of behavior is to recognize it as a strange attractor.

Scaling. Scaling is the primary property distinguishing between a random system and one following a strange attractor. You will remember scaling is the tendency of a complex system to repeat the same patterns in many different parts of the system and at many different levels of scale. If a system is random, then you will find no similarities between behavior in one part of a system and another. Individual behavior is ruled by chance, and the behavior of the whole exhibits no coherent pattern. No one part of the system bears any resemblance to any other part of the system. On the other hand, a complex system under the sway of a strange attractor will exhibit scaling. Though the times and places might vary and the individual characteristics of system parts will be diverse, self-similarity will emerge across the system.

Consider a case of an authentic customer service culture. There might be a great diversity of individual behaviors because every interaction with a customer is unique. If this system were random, you could make no general statements about the kind or quality of service provided to customers. Each employee would follow his or her own whims, and, as a result, no pattern of customer service would emerge. On the other hand, if this system follows a strange attractor, the diversity will remain high, but similarities across the system will be recognizable. Customers will be satisfied consistently. Employees will honestly inquire about the customer's needs and work within the known and allowable options to satisfy the customer in a creative way. Self-similarity, in the midst of diversity, will prevail when a system is under the sway of a strange attractor.

Evolution. Dynamic complex systems evolve. The evolutionary path begins with a point attractor, in which the system moves in predictable ways toward a single outcome. Over time, the system moves into a periodic attractor regime. At first the system oscillates between two possible options. The split of a system from a single to a two-point attractor is called a bifurcation. Later, each of the two options bifurcates again, and the system oscillates among four options. This process is repeated until the system moves among eight then sixteen possible states. When the system is periodically moving among sixteen possibilities, it becomes highly unstable and moves into the next evolutionary step: the strange attractor. Some chaos theorists call this step the "edge of chaos." The system, which in the recent past was predictably point or period controlled, moves into a regime in which neither point nor period appears to play a role. The system's behavior appears random, but it still carries the underlying patterns of highly interdependent behaviors. This evolutionary process from point through periodic to strange attractors is called the period-doubling, or bifurcation, path to chaos.

The development of computing hardware is a good example of the period-doubling path to chaos. At one time the business world depended on monolithic mainframe computers, a single point. Then the period-doubling process began and the field bifurcated into mainframe and personal computers. From that point, personal computers split into Macintosh™ and IBM models, then into business and home models, then networked or not. On the mainframe side, the splits happened as well—supercomputers and regular mainframes, business and scientific applications, networked or not.

Today we have passed the edge of chaos and the emerging strange attractor is known as the massively parallel, client server environment. The old distinctions based on size, speed and location of computing hardware are no longer relevant for the user. Rather, the focus is on business tasks to be performed using all the capabilities of all the machines in the accessible network. This transition has been fairly easy for users, but quite difficult for many technical professionals. They expect a periodic pattern in which the hardware platforms are clearly distinguishable. What they get is an interdependent whole in which patterns are difficult to perceive. They still distinguish logically and physically among the hardware categories of the past and fail to understand the business functions driving system design under the strange attractor.

At the same time the hardware base was evolving, parallel evolutions were affecting software design and the business environment. They too moved through period-doubling steps and crossed the edge of chaos. Their strange attractors now affect and are affected by the one emergent pattern of the hardware evolution. This explains why client server technology reaches across all of the traditional system boundaries of computer science.

In other business and organizational situations, you will note the same period-doubling behavior and evolution from simple, single point attractors toward complex strange attractors.

Change. It is one thing to describe the current attractor regime of a system. After you understand the attractor, you can select specific actions to work in consonance with the existing attractor. The next section provides advice for identifying and working within each attractor regime.

The real management challenge comes, however, when a system should be moved out of its current attractor regime and into another. This need might arise when:

- A company is unduly focused on internal policies and procedures, and it needs to focus some energy on customers.
- A project team is constantly drawn into strange attractor issues driven by the entire corporation so the group cannot complete their team tasks.
- A corporate culture based on one strange attractor is not adaptive to environmental, economic or technological shifts.

Scientific data has shown that changing attractors is not an easy, intuitive task. It requires two complementary actions.

- A seed for the new attractor must be created outside the existing attractor regime.
- Random noise or shocks must be introduced into the existing attractor regime.

Neither of these steps will work alone. If the new attractor is created, without disrupting the old one, the overall system behavior will not automatically restructure to match the new, replacement regime. The system will be stressed because of differences in the two attractors. If the old attractor is disrupted without a suggestion of a new pattern, then the interdependencies supporting the old attractor will become even stronger than they were before. If the noise or shocks to the system are not random, the system develops effective resistance mechanisms to deal with the disruptions and maintain its familiar structure.

Current organizational development practices dealing with change frequently fail because they use constant or regularly periodic shocks to try to disrupt the current structure; they fail to establish or communicate the characteristics of the new regime; or they refuse to disrupt the existing patterns effectively. In the next section, we will discuss ways in which management can work within the natural features of attractors to induce systemic change in an organization.

What Can I Do About Attractors?

So far in this chapter, we have discussed three patterns you can use to describe the overall behavior of a system—point, periodic and strange attractors. We have looked closely at the characteristics of strange attractors to explain why they are so difficult to recognize, manage and change. In this section, we will outline ways you, as a practicing manager, can use this understanding of attractors to your benefit and to the benefit of your organization.

Hints About Attractors

Though no one can control the behavior of a system as a whole, there are specific things you can do as a manager to help your group recognize and adjust to attractor regimes that are best suited to your work and environment.

Allow attractors to develop over time. An attractor emerges from the behaviors of individual components of the system. It cannot be established by fiat. This explains why new mission or vision statements coming down from management on stone tablets are rarely effective. If the statements don't reflect the current complex interdependencies of the system, then they will be (and be seen as) irrelevancies. As a manager, you are one of the components of the system, so your actions contribute to the kind and strength of the attractor, but you cannot make unilateral decisions to adjust or change the rest of the system immediately. Take some action, watch the system's response, then take another action. It will require time for the system to converge on the pattern you envision.

For this reason, you must stay conscious of patterns of behavior over time. Accumulated history will give you hints about how your organization responds in times of stress and in times of relative calm. In the context of history, you can use positive and negative feedback mechanisms to shape individual behavior consistently. Ultimately, the changes in individual behavior will be reflected in changes in overall patterns as they shift to new attractor regimes.

Observe, evaluate and influence. The trick to dealing with attractors is to recognize them when they emerge, evaluate their efficacy with regard to your objectives and to intervene in the system to influence the future paths of the attractor regimes.

To observe, you will collect data continually about the behavior of persons within your organization. What appear to be their motivations? What self-similarity do you observe? How do different parts of your organization respond to the same stimulus? To different stimuli? Look for markers of each type of attractor.

As you begin to recognize attractors in your organization, you will see some as effective and others not. Of course, not all organizational problems are related to attractors, but many are. Every organizational effort will be most productive if its attractor matches the task at hand. When is each attractor type most effective?

After you have described the reigning attractor for yourself and defined the regime that is best suited to the task at hand, you can take action to begin to adjust or adapt the system to a new attractor.

The tables that follow provide tips to help you recognize, evaluate the value of and intervene to influence each of the three types of attractors.

Point Attractor

Suspect one when you see	<ul><li>general agreement on major issues</li><li>repeated phrases or words across the organization</li><li>strong distinction between employees who seek an end and those who have already attained the same end</li><li>single-mindedness among management or employees</li></ul>
They are most effective when	<ul><li>a project must be completed in a short time or with a small budget</li><li>the boundaries for the project are clear and distinct</li><li>the task will either be performed only once or it will be a single focal point for the whole organization</li><li>a small number of persons will be involved</li><li>the employees prefer task orientation and closure</li><li>the system is under extreme stress</li><li>the participants are relatively homogeneous in their backgrounds and interests</li><li>the organization has clarity around the focus and mixed messages are few and far between</li></ul>
To influence one, you can	<ul><li>keep the team small</li><li>have regular and frequent status meetings with fixed agendas</li><li>be explicit about production, time and budget goals</li><li>provide explicit reinforcement for completion</li><li>reduce discussions about options that are not related to the point attractor</li><li>establish measurements and procedures that focus on the point</li><li>close external boundaries as much as possible to reduce distraction</li></ul>

113

Periodic Attractor

Suspect one when you see	• seasonal changes in activity • emotional reactions that vary greatly over time • problems that are critical at one time and less so another • regular variations in financial or process measurements
They are most effective when	• a project or task will be completed repeatedly • external forces on the system are minimal • the work itself or customer requirements are cyclical in nature • options are available for documenting standard procedures to encourage consistency from one cycle to another • the system is not under extreme stress so stages in the cycle can be reinforced • compensation structures depend on process completion, rather than the passage of time • employee skills and interests are varied, but the overall goal is common
To influence one, you can	• celebrate both beginnings and endings • establish rituals that will be repeated on a regular basis • be explicit about the steps required to meet goals • communicate about history of completed cycles • hold regularly scheduled meetings • retain persons on the team who have "been through this before" • perform post-project review meetings to enhance learning between cycles • allow the calendar to determine payment schedules, meeting dates and project activities • make and stick to schedules for activities

Strange Attractor

Suspect one when you see	• behaviour that is repeated, but not exactly and not at regular time intervals • high levels of self-similarity in emotional reaction, procedures or processes • concerns that reach beyond the usual boundaries of your work • multiple inputs, outputs or options for action in work processes • History of increasing complexity
They are most effective when	• creativity is critical to the task • the environment changes continually • personnel reflect a high degree of diversity • redundancy is minimal in products or projects • employees enjoy wide range of freedom • project teams are established on an *ad hoc* basis • time constraints are not rigid
To influence one, you can	• keep boundaries open • encourage individuals to express their personal opinions • be explicit about results, but don't define procedures • hold meetings on an as-needed basis • cross boundaries in selecting persons to serve on a team • provide frequent reinforcement for contributions • help team members be cognizant of self-similarities • introduce random noise into the system in the form of information distribution, unexpected resources or the occasional butterfly effect

Participate in the current attractor. Even if you think the current attractor is not the most efficient, you must participate in it anyway. For one thing, your own frustration level will be lower if you are "moving with the current." For another thing, you can only influence the attractor to change by recognizing its strengths and limitations and working within the existing pattern to establish the new one.

Recognize attractor frustration. When the individual members of a group are working under the influence of different attractors, the result is frustration for all. For example, problems will surely arise when a team leader is focused on accomplishing a task within schedule and budget, and other team members are concerned about completing their monthly reports or understanding the overall economic performance of the company. Such differences in perspective are inevitable, but if boundaries are clearly laid out most people are willing to forgo their own attractors for the purposes of a temporary assignment or a clearly stated objective. Consciousness of the variety of attractors at work will alleviate the friction and improve productivity.

Watch for attractor changes. You can use your role as leader to influence the shift of attractors, but even without your intervention, attractor regimes will shift. External factors, new issues or concerns or changes in management can cause an attractor regime to change dramatically. Any discontinuity in operating procedures, personnel or external circumstances might signal a changing attractor regime. When the shift happens, the rules for productive behavior change drastically and immediately. Be prepared and help employees be prepared to shift their behaviors to match the new attractor.

Establish a combination of attractors. A single-attractor system is vulnerable in times of change. A single point attractor leaves an organization without a focal point when the point is reached. Synchronized periodic attractors exhaust human and other resources at times of high activity and underutilize those resources at other points in the cycle. Strict focus on strange attractors might lead an organization toward divergent creativity without a sense of history or stability.

The key to success is to be sure your organization incorporates multiple attractors and to ensure the attractors harmonize with each other according to the shifting needs of the corporation its employees and customers. Within a healthy, complex organization, point attractors provide a sense of accomplishment, periodic attractors a sense of history and stability and strange attractors an opportunity for creativity and adaptation. Each serves its purpose, and all work together to form a dynamic and adaptive pattern of macroscopic behavior.

Attractors: An Example

In an attempt to improve services and decrease costs, a Canadian province decided to merge all of its departments that supported delivery of compensation to injured workers. Historically the three departments had functioned independently, and their merger caused massive headaches for all concerned. The three included

- An academic institute that pursued a variety of research areas related to worker's compensation
- A network of clinics that delivered services to injured workers
- An administrative council that provided grant monies to academics and practitioners who identified areas of research that would provide potential benefits to the system as a whole

Under the new organization, the administrative council was named as the governing board over both the academic institute and the clinics. They were given the charter to coordinate the activities of the whole department with the two-fold purpose of improving services and decreasing costs. As they established their internal organization, the council divided into two groups. One was to focus on services and the other on fiscal accountability. The new structure created two point attractor systems. The goals were clear, the stress was great and the council believed the point attractor model was the only one to allow them to meet their goals within the time and budget set by the legislature.

The council decided to establish its relationship with the clinics before addressing the academic component of the division. This decision resulted from two factors. First, most of the members of the council were, themselves, practitioners. They logically believed their communications with the clinics would be easier and more productive. Second, the clinics represented the greater financial burden, so it seemed most efficient to begin cost-cutting measures there. Friction developed between the council and the clinics from the first day.

The clinics were loosely associated with each other. They were truly a network in which each clinic maintained its own operating procedures and standards of performance. Their association consisted of informal communications among clinic staff and formal meetings of clinic directors, which took place semi-annually. This network structure functioned effectively within a strange attractor. Each clinic responded to the needs of its local community and, within each clinic, the counselors responded to the needs of individual patients. Though the clinics did complete regular financial reports and participate in the directors' meetings, they functioned otherwise within the domain of the self-similar, but uncontrolled, strange attractor.

When the council arrived with its two well defined point attractors, the clinics ignored them. No matter how loudly the council explained their needs to control costs and increase quality, the clinics refused to budge. Their current levels of services and costs were effectively reinforced by their customers and by other clinics. There was, in effect, a stalemate.

One of the few beliefs held in common between the council and the network was a distrust of the academic arm of the department. Both saw the university researchers as unresponsive to and disconnected from the problems of the real world. As the council realized the clinics were resistant to the changes they suggested, a new strategy emerged. If both the clinics and the council could work together to change the behavior of the academics, then, perhaps, they would work together on other issues as well. The point attractors and the strange attractors laid siege to the ivy-covered walls in an effort to reform the academic researchers.

The research division was structured as a consortium. The full-time director of the consortium was a research physician who had little experience outside the laboratory. She coordinated the activities of approximately fifty researchers in fields as diverse as medical ethics, economics, law, sociology, social work, medicine and ergonomics. The researchers taught in academic settings and performed their research in laboratories or libraries located in six major universities scattered across the province. The consortium never met as a group. The director communicated with the researchers about their grant funding, and the individual researchers communicated with each other, when necessary, over the Internet. The consortium group, as a whole, was so loosely structured it represented no overall pattern of behavior.

The effective attractor for the group was based, instead, on the standard approaches of academic research, which are periodic in nature. The work of the researchers was based on the cyclical structure of the scientific method and on the periodic publication of scholarly research journals. Most of the researchers were involved in longitudinal population studies requiring years of data collection before completion. The journals usually required a year from the time a paper was submitted until it passed a peer review and was placed in print. The academics were so concerned about obtaining professional recognition for their own findings they refused to share information with the clinics or the council until after their papers appeared in print. These two long periodic attractors—scientific method and academic publishing calendars—determined the pace and structure of the primary research work.

As you can imagine, communication was extremely difficult when the point attractor council and the strange attractor clinics approached the periodic attractor of the consortium director. Each struggled with the basic assumptions of all of the others and over a period of months, little headway was made. Each group thought the other groups were becoming more stubborn by the day, and they were. Time pressures, scheduled deadlines and media attention all served to exacerbate the situation.

Finally, representatives from all of the groups sat down during a three-day meeting to try to resolve the issues. Up until then, the groups had been unaware of their own conflicting attractors. As they analyzed their attractor regimes, they began to realize the source of their discomfort and dysfunction. After diagnosing the problem in terms of competing attractors, they began to discuss what an appropriate attractor regime might be for the department as a whole.

Their conclusion was to define a common periodic attractor to meet the needs of the legislature who chartered and would continue to fund them. The legislature had asked for quarterly reports of progress, so it appeared to be the appropriate cycle of work for the department as a whole. Each of the three groups adapted some of their behaviors to coincide with the quarterly periodic cycle. The council broke its single-minded goals into quarterly objectives to synchronize with the changes taking place in the network and in the consortium. The network identified specific measures of quality and cost and reported their performance against those measures on a quarterly basis. The consortium agreed to sponsor a quarterly workshop for the clinic directors if the directors promised to treat the findings as preliminary and proprietary. During the workshop, members of the consortium would present their papers and discuss practical applications and implications of their findings. Summary minutes of the workshops would be submitted as part of the council's quarterly report to the legislature.

Individually, these changes look like minor adjustments to the separate parts of the system, but they made tremendous changes in the performance of the department as a whole. The increased communication and opportunities for transforming feedback allowed the three groups to maintain their autonomy and to adapt to the contributions of the others. It established a whole that was greater than the sum of the parts.

Attractor analysis can help you describe and resolve issues that appear to be massively complex. Each situation is unique, but patterns of interaction can provide cues for effective intervention. Attractors provide just such informative patterns.

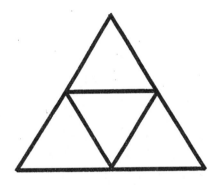

Chapter 6
Self-Organization

Chapter 6
Self-Organization

Tragic heroes in Greek drama share one fatal flaw. They call it *hubris*, the belief that a human being can control fate. Sometimes it is translated as pride, other times it is translated as arrogance, but the message is still the same. Human beings should know what they can and cannot control. To attempt to control the unknowable is a tragic flaw.

Modern science provides mankind with mathematical and mechanical tools that allow us to control many things, but it also leads us to forget about the aspects of nature beyond our control. In the late 20th century, science has begun to confront the uncontrollable aspects of nature. In meteorology, high energy physics, biology and social sciences, researchers acknowledge some phenomena beyond the control of mathematics and mechanics. Science gives up the role of objective controller of nature. It seeks only to understand and explain.

Management theory has also adopted the *hubris* of modern science. The manager is seen as the almost-invisible hand to structure, motivate and execute visions for the future. A manager is seen as an organizational engineer who defines the components of a system, determines how the components work together and controls the mechanics of change in the work place. This modern version of *hubris* has resulted in a variety of personal and organizational tragedies. Managers accept the assumption that they should be able to control all activities and outcomes in their domains, but they are constantly out of control. Organizations invest tremendous resources to formalize all processes and procedures, hoping to ensure optimal outcomes, but the unexpected intervenes to derail the best laid plans. Economic control over natural resources deplete the world's wealth and generate unlimited waste that must be destroyed or stored for eons. International conflicts arise over control of governments and nations, when none of the combatants has any hope of truly controlling the future.

The sciences of complex systems provides an alternative to this hopeless drive to control everything. By recognizing the complicated interdependencies of complex systems and acknowledging that the leader is not objective, the management sciences of the late 20th century allow leaders to participate in, but not control, their environments.

This chapter presents the complex system's alternative to the *hubris* of the modern manager. This factor affecting the behavior of complex systems is called self-organization.

What Is Self-Organization?

Many complex systems exist at equilibrium in a kind of dynamic balance of change and stability. The differences among the parts are not too great, and the flow of information and energy from one part of the system to another is continuous and smooth. If they are healthy, productive complex systems, however, they are also connected to the world beyond their own boundaries. External sources of information, materials and energy are available to the system and influence its internal dynamics. As these external forces shift, they might disrupt the internal equilibrium of the system. When the balance of a complex system is disturbed, the components of the system adapt to their distressed environment. Differences among components are exaggerated. Flow of resources within the system becomes turbulent. The system is pushed away from equilibrium and into what is called a far-from-equilibrium state.

When a complex, interdependent system is pushed far enough from equilibrium, a strange and amazing thing happens. Without an obvious plan and without the control of any single individual component, the system spontaneously reorganizes itself. This process is called self-organization.

Examples of this phenomenon abound in nature. You begin to play chess by learning the rules and the moves. At some point you have an "aha" experience and grasp the meaning of strategy. The details of the play have self-organized into a holistic understanding of the game. An egg and sperm join and the single cell begins to divide. After four divisions, the cells differentiate themselves into the self-organized structures to sustain life. A specific combination of chemicals at a given temperature and rate of flow will spontaneously switch colors at a regular rate of speed. This "chemical clock" demonstrates how the molecules organize themselves into a single, concerted behavior. The slime mold functions as any other mold when nutrients are easily available. When the food runs out, however, a chemical message passes through the colony and all of the cells move toward a single point. The point at which the cells converge becomes the base for a flower-like growth that enables the reproductive cycle and distributes the colony into a wider area where food might be more plentiful.

In all these cases, a complex, interdependent system responds to its environment by absorbing information, energy or material. The new resources stress the system by moving it out of its equilibrium state. Then, the system spontaneously generates a new structure. It self-organizes.

None of these cases involves an all-seeing manager who is in charge of the change. None is driven by a visible hand. None is controlled by an individual source of information or energy. All of them result from high interdependency among components of a complex system and transforming feedback loops to allow changes in one part of the system to influence adaptation in other parts of the system. This is self-organization in complex systems.

Self-Organization in Human Systems

What does this mean for organizations? If a human organization is a complex, highly interdependent system, then it will organize itself when it is pushed out of its comfortable equilibrium conditions. You have probably observed situations in which this phenomenon occurred. Factions and cliques form during early phases of massive organizational change. Informal communication networks are established quickly when familiar, formal channels are disrupted by reorganizations. A collection of disenchanted persons becomes a mob that appears to have a mind of its own. Children playing in a room will construct group activities and games without any adult intervention. Groups of consumers come together to form activist coalitions. A spokesperson or representative of a group will arise by consensus among all participants. New teams establish standing jokes and short-hand vocabulary to help them identify with each other as individuals and with the group as a whole.

In all of these situations the organization experiences a drastic change in structure without any conscious or controlling intervention on the part of formal authorities. In some cases the resulting organization matches the formal objectives of the group, in other cases the new structures interfere with the group's work. From the formal manager's perspective some are constructive and others are destructive. Later in the chapter we will discuss how you can participate to help influence the self-organizations of your group. But first, we need to define the conditions required for self-organization to take place.

128

Preconditions for Self-Organization

Not all complex systems self-organize, and not all organizational structures result solely from self-organization. Within a system with a potential for self-organization, a formal, conscious, centrally controlled organizational change can also be implemented. In this section we will discuss the fundamental conditions necessary to support self-organization. Those conditions are differentiation and transforming feedback loops.

Differentiation. The differences in a system provide the opportunity for dynamic motion and change. As we pointed out when we discussed boundaries in Chapter 2, difference provides the impetus for movement and change within the system. If a system is homogeneous, there is no potential for movement or change. No impetus exists for adaptation or adjustment.

Differentiation in a complex system works like potential energy in a mechanical system or voltage in an electrical system. A ball will not spontaneously roll anywhere if it is on a level surface. Only differences in height provide the potential energy to cause the ball to move. Electrical current will not flow if there is no potential difference between two points in a circuit. A dead battery, which has lost the ability to sustain a difference in voltage from one pole to another, will not generate any electrical current. In the same way, a complex system in which all of the components are the same cannot generate a flow of material, information or energy. In essence, the system is dead.

An organizational system in which all the parts are the same also loses its potential for change. This is the reason why cultural diversity in corporations is more than just political correctness; it is a definite competitive advantage. It also explains why authentic attachment to customers and their unique perceptions can play a critical role in transforming an organization. The customers' expectations and visions are different from those of employees, and this difference can generate movement and adaptation of the organization as a whole.

Differentiation alone, however, is not sufficient to engender self-organization. The system must also provide pathways for communication and change among its component parts. Transforming feedback loops provide the route for communication.

Transforming feedback. As you learned in Chapter 3 transforming feedback forms the communication links between component parts of the system. They allow change in one part of the system to be reflected in all other parts of the system.

Not all complex systems include transforming feedback loops. In some systems, component parts are not in communication at all. They might be separated by an impermeable boundary, or they might be so far removed from each other they have no opportunity to interact. When transforming feedback loops are absent from the complex system, self-organization is not possible.

When the system incorporates critical differences and when transforming feedback loops join the different parts of the system into a communicating whole, then self-organization is a natural phenomenon in the complex system.

The next section provides a graphic tool to help evaluate the potential for self-organization in a complex adaptive system.

Self-Organization Matrix

When working within a complex system you will observe a complicated combination of differentiating factors and a wide assortment of feedback loops. The table below will help you categorize, describe and influence change in your complex organization.

	High Differentiation	Low Differentiation
Active Feedback	Self-organization	Reinforcement & revitalization
No Feedback	Unresolved conflict	Organizational rest

Let's consider each of these quadrants and the kinds of behavior you can expect when an organization varies its degree of differentiation and its available feedback mechanisms.

Self-organization. The self-organization quadrant represents high differentiation and high levels of transfer within the system. In this situation, you will observe the behavior of a far-from-equilibrium system. Lines of information transfer and potential transformation appear between areas of highly differentiated components of the system. These pathways, similar to the convection currents in heating water, will form self-organized structures. Examples of behaviors in such organizations include the following.

- Constructive conversations among marketing, distribution, product development, administrative and sales departments in which each group learns from and adapts to the concerns and knowledge of the others.
- A corporate penchant for learning from each new experience through the explicit discussion of mental models of the phenomenon and attention to patterns of distinction that emerge from the whole.

- Effective and swift response to shifts in customer, economic and technological environments that are made possible by integrating new information into the problem solving strategies of the organiza-
- tion. Authentic empowerment for all levels of employees when the perspectives of each are integrated into the organization's view of itself and its environment.

Though groups in this quadrant will generate self-organizing structures, it is not always the "best" situation for a constructive, well-functioning team. When groups exist too long in this quadrant, they exhibit some negative behaviors.

- Signs of stress and exhaustion when constant differentiation and continuous transformation are required.
- Preoccupation with change and negotiation regarding issues that might not have substantial effect on the performance of the primary business functions of the organization.
- Ineffective allocation of resources and organizational focus when irrelevant differentiations dominate the attention of the group for extended periods of time.

Complex patterns of behavior arise in organizations when they exhibit high differentiation and high transfer. The differentiation provides vivid patterns, and the transfer allows for dynamic adaptation to new system inputs. The complex dynamics of such interactions allow for growth and change. The emergent structures truly dissipate entropy (the measure of disorder) and give new meaning and context to information and individual roles within the community.

Reinforcement and revitalization. The reinforcement and revitalization quadrant represents a system in which differentiation is low, but transfer of information is high. In such a system, one might recognize high levels of energy, but low levels of meaningful information. Such a system is close to equilibrium, but the equilibrium is dynamic because information or energy is continually transferred among the parts of the system. These systems can be characterized as "preaching to the converted," perpetuating common and popular procedures and beliefs and generating high levels of cohesion and loyalty. Cliches emerge that give the appearance of understanding without the substance. Examples of behaviors in such organizations include the following.

- Business or professional teams that function as fraternities of like-minded persons who personally identify with each other.
- Groups that exhibit strong aversion to and difficulty in resolving conflict because the expectation of differentiation as a threat to the ongoing success of the group.
- Management teams that are dominated by autocratic leaders who reinforce employees who mimic the leader's ideas, attitudes and values.
- High technology companies that lose sight of customer needs and competitor's advances because technical personnel reflect each others' common view of the technical aspects of their products or services.

In contrast to the self-organization quadrant, this might appear to be a negative situation, but such is not always the case. In some situations, high transfer and small differentiation might be a constructive situation for a group.

- A team needs to celebrate its group accomplishments. Commonality of the group should be communicated and reinforced.
- New members of a team need to be initiated into the culture. Initiation must take the form of active communication about the shared processes, history and activities of the group.
- Basic vision, mission and values of a group should be reinforced during turbulent economic times.

These low differentiation, high transfer systems encompass a high degree of redundancy. Their patterns of behavior are subtle and consistent. Security is provided in such organizations to meet the emotional and social needs of many members of the community.

Unresolved conflict. The unresolved conflict quadrant describes systems in which differentiation is high, but transfer of information is low. In such human systems, one would recognize a high degree of anger or frustration. The differences are so great and the communication barriers are so strong no one component of the system can affect another component. These systems are similar to hot and cold water separated by a thermal shield. They contain a great amount of potential energy, but the energy is locked within impermeable boundaries. Examples of behaviors in such organizations include the following.

- Harassment based on gender or culture when a dominant group acts on its ignorant biases about another group.
- Fear of an autocratic leader in which frustration or disappointment with management engender only anger and resistance from employees.
- The traditional distrust between product development and product support personnel when their objectives and procedures seem disassociated from each other.

Again, this quadrant is not always a bad situation in an organization. A leader might guide a group into this quadrant in the following circumstances.

- Upper level management becomes aware of a long-term threat to the organization, and they need time to evaluate the implications before disturbing employees.
- Technical experts are asked to investigate a new and complex product line, which might or might not move into design and production processes.
- One employee receives a negative performance appraisal and communication between the employee and others should be kept to a minimum.

134

The hierarchical structure of traditional organizations encourages such high differentiation, low transfer cultures. Interdependencies of individuals and functions are not explicit or valued within the organization. The organization as a whole cannot leverage the strengths of one subgroup against those of another because networks and habits are not established to support communication across lines of diversification.

Organizational rest. The organizational rest quadrant describes systems in which there is low differentiation and low transfer. These systems are closest to equilibrium, and their information content is not high. In such organizational relationships, the following behaviors can be observed.

* Lots of smiling and nodding, but little talk in meetings because everyone already knows what everyone else knows.
* High resistance to change because the friction of differentiation is abhorrent and communication networks do not support transformation.
* High reliability and low tendency to error because processes and functions are redundant, well practiced and stable.

In some circumstances, this is the most effective situation for a company.

* When processes or procedures should be repeated with little variation, differentiation is truly a threat to consistency and quality.
* Individuals in an organization are threatened by burnout, and they need time to adjust their personal visions and tasks to meet the expectations of the group.
* Resources, time and energy must be conserved for future activities.

If left to themselves, these systems approach death of minimum differentiation, as differentiation is minimized and transfer is interrupted. Actions appear to be random and insensitive to changes in external conditions. All boundaries disappear

135

and the distinctions to drive growth and adaptation cannot be discerned.

In the descriptions above, I have explained the quadrants as characteristic of an organization as a whole. Every organization has its most common dynamic—some revel in the vibrancy of far-from-equilibrium status, and others seek the security of total equilibrium. But this simple, two-dimensional description is not sufficient to describe the complexity of organizational behavior. At least three variations complicate the description of any real, functioning group.

First, parts within the organizational whole might vary in their most common status. For example, product development personnel might function effectively within the far-from-equilibrium status, while accounting might seek the lowest levels of differentiation and transfer. These distinctions among subgroups must be identified and scaled within the context of the organization as a whole.

Second, through time, an organization moves from one to another of the quadrants. During start-up or periods of extreme turbulence, an organization might do well as it leverages differentiation and transfer of information. But, as it reaches maturity or stable balance with its environment, the same organization might move toward lower differentiation and transfer.

Organizations also might move from one quadrant to another depending on the type of issue addressed. During the same meeting, a team might have low differentiation and transfer when discussing a week's progress against a goal and switch to high differentiation and transfer when the topic turns to basic design or customer service issues.

Finally, every organization encompasses many levels of differentiation and feedback loops at any given time. Individuals, teams, departments and the organization as a whole wrestle with differences and feedback loops too numerous to count. This same dynamic of activity within and motion between quadrants of the matrix is repeated for every difference and every level of the organization.

When the preconditions for self-organization are present, the system spontaneously reorganizes itself. This process is not painless or automatic, but it is spontaneous and is controlled by the complex interdependencies inherent in the system not by the manger. Any manager who discounts the innate ability of the system to self-organize or believes he or she is in control of all of the organizational processes within the system risks the ancient Greek weakness of *hubris* and the personal and organizational tragedy engendered by it.

What Can I Do about Self-Organization?

You can use the self-organization matrix to assess the current state of your organization with regard to its potential for self organization, but you must also be aware of tactics you can use to establish the preconditions for self-organization and to participate actively to help shape the outcome of self-organization. The next section will provide pragmatic tactics to help manage self-organization in your group.

Managing Self-Organization

You cannot control self-organization, but you can influence it as an active participant in the system if you are sensitive to the system and consider some of the tips described below.

Encourage differentiation. Self-organization depends on sufficient variation in the system. As a manager, you can take specific steps to encourage and reward differences in points of view, skills and working styles. Consider the following steps.

137

Identify constructive boundaries of distinction within the group. Not all distinctions are equally important to the completion of a particular task. For example, gender distinctions should be irrelevant when a question of technical design is at issue. You can focus a group on relevant distinctions and move their attentions away from irrelevant or destructive differentiations for a specific task.

Use difference questioning to encourage constructive differentiation in the group. Difference questioning allows a group to identify and make explicit their differentiations. This can be a first step toward moving from low or unconscious differentiation into a realm of explicit and constructive differentiation.

Recognize and reward diversity within the group. In the rush of daily activities, it is easy to think of differences as distractions from efficient work, but this approach will tend to decrease the amount of creative diversity within the group. Seek and give credence to opinions different from your own. Take time to recognize and reward constructive differences to encourage their continued contributions to the group's activities.

Select employees who represent diverse backgrounds and cultures. Avoid the tendency to hire persons who reflect your skills and style of work. Seek to incorporate others whose perceptions and perspectives add diversity to your group's functioning.

Spend time with employees individually to allow them to define their own differences. Until a culture is established to value diversity, focus individual attention on employees. They will be more likely to discuss their unique insights with you in private, then, with their permission, you can share their perceptions with the group and discuss the value of the new perspectives openly.

Encourage transforming feedback. Chapter 3 provides detailed information about establishing transforming feedback loops, but consider the following suggestions as well.

Provide a safe environment in which participants can express their ideas and opinions. Security is a major factor in a group's tendency to low transfer. Structure your environment and procedures to provide security through other measures, so employees will feel safe voicing divergent opinions.

Encourage communication around issues that are highly leveraged for the work of the group. Focus transfer of information on issues that are significant to the group and facilitate discussions to focus on ideas, not personalities.

Begin communication on issues for which there is not a high degree of diversity or emotional involvement among the group. In this way, you can help the group develop the habits of constructive conversation and transformation before turning the group's focus to significant or highly contentious issues.

Listen, listen, listen to model effective transfer and transformation. Constructive transformation requires every member of a group to speak and to listen intently.

Provide a variety of venues for informal communication to take place. Information transfer can become the way of life for an organization, but this seldom happens when significant conversation is restricted to formal environments. You can structure a work place in which information transfer is encouraged in all places at all times.

Encourage public discussion, as opposed to private discussion of issues. When an idea arises in private, find ways to air it in public. The differentiation is not valuable to the organization as a whole unless it is accessible to all.

Reinforce efforts to communicate. Use any option at your disposal to encourage and reward effective transfer of information and transformation.

Remove constraints on self-organization. Most organizations, with their focus on negative feedback and control mechanisms, interrupt or distort the natural tendency to self-organization. Observe your own environment, identify systemic factors that counteract self-organization and work to remove them.

Reinforce trends toward constructive self-organization. One of your primary tasks as a manager of self-organization is to observe the changes as they begin to take place and to use reinforcement techniques to encourage constructive patterns. Keep your eyes open for signs of impending self-organization and reward the most adaptive movements.

Stay out of the way. Finally, any action you take as a manager will be seen as "other organization" rather than self-organization. Whether you relish the idea or not, you are seen as an external force in the dynamics of your work group. For this reason, you must stay out of the way as self-organization gains momentum. Early in the process you can intervene with positive feedback, but late in the process your interventions will only distort the natural flow of the self-organization process.

Self-Organization: An Example

As you focus your attention on the potential for self-organization, you will notice innumerable individual incidents in which it is a deciding factor. In the following example, however, the self-organization of a group of international scientists changed the face of technology integration on a world-wide scale. This case study is the result of the research of Dr. Ralph Stacey, Professor of Management Hertfordshire University Business School in the United Kingdom. Dr. Stacey presented the case in a paper for the Chaos and Society Conference, held in Hull, Quebec, Canada in June of 1994.

Beginning soon after the Second World War, individuals in various European countries assumed roles of formal and informal advisors to their governments on matters of science and technology applications. These individuals became acquainted with each other through chance encounters and regular attendance at international scientific conferences. They soon found they were able to provide information and support to each other and to their respective governmental agencies. An informal club of technical advisors emerged from this group of cooperating individuals. And the club ultimately developed into an official body with a chairman and executive committee: the European Technology Liaison Committee. By 1975, the committee had evolved into an institutionalized collection of formally appointed representatives of government departments, and it was called the ETTC.

In the 1980s, Jose Fonesca became the Executive Secretary of ETTC and Head of the Technology Division. His primary concern was to break down the technology division between developed and developing countries and to establish team efforts and quality controls in the integration of technology in developing nations. He tried to pursue these ends through formal channels, establishing a review group of nominated ETTC members to identify priority areas for future technology efforts. The review group produced a report of their findings and recommendations in 1984. Formal efforts at implementation of the group's recommendations failed because of a combination of political, economic and logistical pressures.

After the publication of the formal report, Fonseca hired a temporary aide, named Johnson, to implement the recommendations contained in the report. When Johnson came on board, he discovered a workshop had been scheduled for early 1985, and he was asked to organize it. The purpose of the workshop was to initiate a "management of technology" program. Rather than running a traditional workshop with presentations of formal papers, Johnson decided to create an interactive problem-solving environment for the workshop. He designed and developed an expert system, which he called VALTEC. This tool made it possible for policy makers to ask a set of specific questions about research projects and be guided by the system in making judgments about relative priorities.

Fifty people attended the conference. They were a mixture of scientists and civil service bureaucrats from a variety of levels within their departments. When Johnson presented them with VALTEC, the attendees decided to divide themselves into teams based on their common sets of problems. The purpose of the teams was to understand how to use VALTEC, to apply it to their own problems, and share their findings and questions with the group.

In this team-based, self-organized environment, the lines between developed and developing nations disappeared. Instead, all participants focused on their common technical concerns and problems. The teamwork brought great success during the workshop and established long-term relationships that continue to support productive work among nations. By early 1987 there were about 80 individuals actively participating in a worldwide network of technology policy makers assisting each other on issues of mutual concern. The network operated on a system of bartering favors. Gradually the group adopted more formal organizational structures until it was decided in early 1994 to set up a limited liability company to run the affairs of the group.

Formal hierarchical power structures did not support the self-organization process of this informal group of individual contributors. Repeatedly the group was forced to search for opportunities and support through informal channels. In spite of a lack of support, the differences in levels of technical expertise and the rewards of cooperation moved the process of self-organization forward.

This story illustrates how feedback loops among groups of different locations, areas of expertise, and experiences can generate a powerful process of self-organization without the assistance of formal organizational structures. As a manager, you to can participate in and work to allow self-organization to shape the future of your organization.

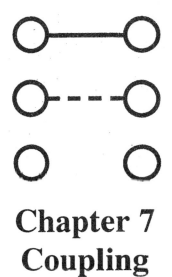

Chapter 7
Coupling

Chapter 7
Coupling

What is the difference between driving a motor boat and sailing?

In a motor boat, you prepare for a trip by checking the engine, the fuel tank, the life jackets and the maps. The focus is on the physical resources within your direct control. You plan a route and a destination, start the engine, put it in gear and steer your predetermined course. You call the trip a success if you arrive at your destination at the expected time with a minimum number of surprises along the way.

Sailing is different. You check the direction and strength of the wind, ensure all of your gear is in order, instruct your crew in their respective responsibilities, agree on methods of communication during the trip. You might have a general idea of destination and estimated time of arrival, but you and the crew know these plans are open to change, based on the variables inherent in the environment. You call the trip a success if you and the crew adjust quickly to the changing circumstances, work together well during the journey and arrive safely at or near your destination within a reasonable time.

Management in a complex system is much more like sailing than driving a motor boat. The captain of the boat focuses on the surroundings, makes adjustments during the trip, constantly communicates with the crew and depends on the individual skills and vigilance of everyone to ensure a safe and enjoyable trip. The pace and path of the voyage are determined more by environmental shifts and forces than by the predetermined plan of the captain.

When working with or within an unpredictable environment, influence, rather than control, determines the process of change and the outcome of the leader's actions. In such situations, the degree of influence exerted by the leader, team members, other stakeholders and the environment determine the efficiency and effectiveness of the group's planning and implementation activities.

Scientists who work with complex systems in the natural world have a special name for influence between the parts of the system. They call it coupling. The parts of the system and their subsystems are coupled together, and the coupling determines the amount and kind of influence each has on others and on the performance of the whole.

In this chapter, we will examine three levels of coupling and investigate how each affects the behavior of the system as a whole. We will also identify some ways the manager can design combinations of coupled relationships to optimize the performance of a group within a shifting environment.

What Is a Couple?

Influence, not control, is the driver in complex adaptive systems. Any two parts of a system that are considered to be related through material or informational resource exchange are coupled together. Scientists define three different types of couples in complex systems: Tight couples, loose couples and uncouples.

Tight Couple

As the name implies, a tight couple indicates a high degree of influence between two parts of the system. You can think of the couple as a special kind of transforming feedback loop in which the transformation resulting from the feedback is immediate and dramatic. (Refer to Chapter 3 for more information about transforming feedback loops.)

When parts of a system are tightly coupled, a small change on one side of the boundary results in an immediate and radical change on the other side of the boundary. Imagine one heavy object and one light object that are free to move independently, but are connected together by a tight spring. If the heavy object is moved slightly, the other might shake uncontrollably or begin to swing wildly. This tight couple not only ensures an immediate response from the other side of the boundary, but it usually indicates the responding action will be unpredictable and potentially uncontrollable.

In real life, you frequently see tight couples between parents and young children in the grocery store. The family moves among the aisles happily, examining products and selecting items they wish to purchase. At some point, the child makes a move or utters a statement leading the parent to believe the child is moving out of control. In response, the parent takes a rigid stand against the offending action or word. The child, responding to the parent's new-found rigidity, begins to test the boundaries by exaggerating the original act or by adding some other offensive behavior. Any small change in either parent or child causes a dramatic and immediate response from the other, and an insignificant, off-hand action fuels an all-out war between the generations.

In a work environment, tight couples can be either constructive or destructive. For example, an abusive boss uses tight couples to maintain a reign of terror over employees. Her every move requires some immediate response on the part of every employee. The most dramatic and immediate response receives rewards, and so every person in the system strives to "improve" their work by responding faster and in more exaggerated ways to future messages from the boss. This tight coupling, which focuses attention on the boss, is disruptive to the other work of the group because it absorbs attention that should be placed on processes or customers. It also results in discontinuity in relationships among other employees and in daily work patterns.

On the other hand, a service business needs to establish tight couples to customers. Such a tight couple allows the service provider to recognize and respond to the needs of each customer immediately and completely. Anything less than a tight couple between an organization and its customers introduces the risk of eroding customer satisfaction and decreasing market share.

Tight Couples

Benefits	Risks
• Real and perceived responsiveness between the parts of the system • Fast change and adaptation in response to messages across boundaries • Sense of security • Clear lines and measures of accountability	• Uncontrolled or uncontrollable change • High stress among workers • Focus on the single tight couple to the exclusion of other system components • Discontinuity of processes and procedures • Inappropriate use of power held by one individual or group • High levels of fear and defensiveness in the group

If you want to gain the benefits and avoid the risks of tight couples, you must consciously select the boundaries across which tight couples are needed and establish a network to balance tight couples with both of the other types—loose and uncouples.

Loose Couples

A loose couple results when a feedback loop allows transformation to be gradual and not radical. Influence is exerted by one part of the system on the other, but the responding transformation is also influenced by other factors in the system. A physical example is two equally heavy objects, free to move, but connected by a weak spring. One object begins to move and the other receives waves of energy that might be too weak to force immediate response. The inertia of the second object, in addition to the motion of the first, influences the change. Ultimately, the second object might begin to sway and its motion will also influence the first. After some time of mutual transformation, the movement of the two objects will be synchronized.

A constructive argument between spouses is a wonderful example of loose coupling. The wife sees a problem in one way. The husband sees the same problem in a different way. When each is able to state his or her position, and the other is able to hear and adapt, the pair is able to determine a solution to meet the needs of both.

In business, you see constructive loose couples in effective cross-functional teamwork. The needs of one department influence the needs of the other, but the responses are neither immediate nor complete. Each department recognizes how the other must also respond to influences beyond the immediate relationship. In other words, this one couple functions in concert with other couples affecting the work of each separate department.

151

In business, you see destructive loose couples when the unscrupulous practices of one employee go unchecked and influence the behavior of others. One influences the other over time, but the influence is a dangerous and potentially destructive one.

Benefits	Risks
• Allow for smooth transition, adaptation or change • Provide common ground for disparate parts of an organization • Evolve over time in response to the complex nature of influences in the system as a whole • Provide continuity in change • Establish opportunities for generating solutions to be more effective than those generated by one system component alone • Provide a sense of control for all parts of the system	• Can be slow to exhibit system change • Demonstrate subtle early signs of change that might be difficult to discern • Make it difficult to locate a single cause for a change • Function within the informal as well as formal lines of communication • Depend on individuals and sub-systems to negotiate the amount and content of change on an ongoing basis • Generate unpredictable outcomes

When mixed with other kinds of couples, it is possible for an organization to optimize planning and implementation by depending on loose organizational couples.

Uncouples

Sometimes, the parts of a complex system become totally dissociated from each other. The result is an uncoupled system. When the parts of the system are uncoupled they have absolutely no influence on each other. Imagine two blocks of wood lying on a table. One can move without affecting the other in any appre-

152

ciable way. The blocks are uncoupled from each other because no influence is discernable between them.

You see uncouples when the members of a family are estranged. The actions and thoughts and environments of one have no affect on the other. The impermeable boundary between the two ensures no communication or coordinated change can occur.

In an organization, we frequently see uncoupling occur between the sales and the research and development departments. Researchers focus on technical possibilities for new products and services, while sales persons focus on current products and customers' immediate needs. Neither communicates with the other. This uncouple frequently results in wasted investment in research and development and damaged customer relationships when new products and services are not influenced by customers' perceived needs.

On the other hand, uncoupling can be an organizational necessity in issues such as information and property security, legal limitations and employee confidentiality. In all of these arenas, it is critical that one part of the organization be set apart from another part. Uncouples in these cases result in responsible management of resources and information.

Uncouples

Benefits	Risks
• Clear lines of accountability and authority • Minimal interdependence • Maximum freedom of action • Increased sense of self-control	• Insensitivity to complex influences within the system • Avoidance of adaptation and change • Unnecessary redundancy • Inefficient use of resources • Fragmentation in mission, vision and structure • Organizational fear and suspicion

As with the other forms of coupling, an uncouple is a powerful tool for good or for ill in any organizational setting. When used wisely, uncouples provide organizational security and stability. When used unwisely, they result in fragmentation and inefficient use of resources.

In the next section, we will demonstrate how you can establish a combination of tight, loose and uncouples to optimize resource and information flow in your organization and to encourage healthy adaptation.

What Can I Do about Coupling?

Every organization is a complex network of systems and subsystems, each of which has established a coupling relationship with all others. As a manager, your first task is to identify the current couples affecting the performance of your group and to evaluate their effectiveness. After you understand the existing network, you can begin to take steps to alter the coupling habits of the organization as required to increase efficiency and enhance adaptability of the parts and the whole.

Managing Coupling

The analysis and design process for adaptation of coupling in an organization includes three steps.

1. Describe the current state of affairs.
2. Evaluate the effectiveness of the current state.
3. Take steps to alter the coupling structures, if necessary.

1. Describe the current state of affairs. How do you find out about the existing couples in your organization? The basic process is simple, but you will need to repeat it for every significant boundary and feedback loop in your current system. You should start by defining the boundaries affecting your working group, as described in Chapter 2. After you have defined all of the boundaries, identify the ones having greatest effect on your work and consider each of them separately. As you consider a boundary, your goal is to determine what kind of a couple exists across the boundary. Use the list of questions below to characterize each boundary.

How often does information flow from one side of the boundary to the other?

- If the information flow happens on a regular basis daily or weekly, the couple is likely to be a tight one.
- If the information flow is infrequent or if the information transfer comes at irregular intervals, then the couple is probably loose.
- If no information flows across the boundary, then the system is probably uncoupled.

What response is expected when information flows across the boundary from one subunit to the other?

- If an immediate and predictable response is expected, then the couple is probably tight.
- If the response allows for a time lag and if the receiver is free to fashion the content of the response, then the couple might be loose.
- If no response is expected, the subsystems are uncoupled.

How many lines of communication cross the boundary?

- Usually, a tight couple will depend on a small number of formal communication links.
- A loose couple usually includes multiple information pathways, many of which will be informal in nature.
- An uncoupled system will include no lines of communication either formal or informal.

What are the relative strengths or powers of the subsystems that meet at this boundary?

- Tight couples usually connect two subsystems that are not equal in strength or power or size. The larger and more powerful component seeks to drive the smaller through a tight couple.
- Loose couples are more likely between subsystems that are commensurate in size or power. Neither has the potential to completely control the other, so they develop mutually influential relationships through loose couples.
- Because the uncoupled subsystems are dissociated from each other, there is usually no restriction on their relative size or strength.

2. Evaluate the effectiveness of the current state. After determining the kind of couple, your next step is to evaluate the behavior around the couple to determine if the existing couple is meeting the needs of the individuals and subsystems involved as well as the needs of the organization.

Consider changing a tight couple if any of the following conditions exists and interrupts constructive work. Leaders are overwhelmed with the all-consuming responsibility for "running the show." Followers are fearful or unempowered within the current structure. Change activities begin with a flurry on both sides of the boundary, but soon dwindle until the state of the whole returns to normal. Large amounts of resources are spent fighting fires identified by the controlling component of the couple. Standard operating procedures are frequently dismissed in times of crisis. You see evidence of inappropriate uncouples of either component from other subunits of the system.

Consider changing a loose couple if any of the following conditions exists and interrupts constructive work. Action, change or adaptation are slower than they should be across the boundary. Accountability is unclear between the subunits of the system. Informal networks work at cross purposes to the formal networks of communication. Destructive adaptation occurs more frequently than constructive adaptation between the subunits.

Consider changing an uncouple if any of the following conditions exists and interrupts constructive work. Information or resources are held by one subsystem and not available to the other. Individuals within the groups fail to recognize the mission, needs or constraints of others within the system. Humorous or prejudicial comments are made in one part about the other part of the system. One subunit exhibits distrust of the other. Redundancy of process or resource allocation between the subunits is unproductive or inefficient. Discussions on either side of the boundary focus on optimizing the part without considering the needs of the whole.

After you have used the preceding questions to determine which boundaries require change, then you are ready to take action to adjust and adapt communication networks to alter the nature of the current coupling relationship in the system.

3. Take steps to alter the coupling structures, if necessary. This step, of course, is the most difficult and ambiguous of the process. Every complex system is unique in its combination of forces and influences. Over time, the system's influences evolve, and the system changes. For this reason, you must experiment with strategies and tactics to determine which are most effective in a given system at a given time.

The suggestions below are intended to provide a range of possible actions to adjust the coupling in a system. As you work

157

to restructure coupling relationships, you must try a tactic, test its effectiveness and try another, if necessary. You should also be creative in looking for ideas and actions you can take to change destructive couples without disrupting the constructive ones.

If you want to create a tight couple, implement tactics to increase dependence of one part of the system on another. Consider the following tactics.

- Institute regularly scheduled communications in the form of meetings or formal memos and communiques.
- Measure performance across the boundary.
- Build explicit accountabilities into the system.
- Increase the frequency of communication.
- Establish, and measure adherence to, standard operating procedures.
- Mix positive and negative feedback to control behavior across the boundary.

If you want to create a loose couple, develop ways to increase interdependency within the system. Consider the following tactics.

- Construct a communications process to support communication on demand, rather than on regularly-schedule agendas.
- Base communication on informal face-to-face encounters, rather than formal meetings or written communiques.
- As a leader, speak less and listen more. As a follower, look for ways to help the leader hear and understand your concerns and constraints.
- Slow down feedback loops to allow for mutual adaptation between encounters.
- Build two-way, authentic communication between the components.
- Establish multiple couples of each component to others within the system.
- Depend on positive feedback to encourage change across the boundary.

If you want to create an uncouple, take actions to decrease interdependence between the parts of the system. Consider the following tactics.

- Make your intentions to uncouple explicit with the other group. If you don't they will continue to try to influence behavior across the boundary.
- Stop all communications going either way across the boundary.
- Be sure everyone in the system is aware of the rationale for uncoupling and knows the strategies to carry it out.
- Set a timeline for establishing a new loose couple, if it seems reasonable to do so.
- Change any internal policies and procedures that currently link the groups across the boundary.

These are just a few of the strategies you can use to change the coupling relationships within your system. As you test these techniques and observe the results keep the following guidelines in mind.

If you change the couple at one boundary, relationships at other boundaries are likely to change as well. For example, developing a tight couple with the customer might cause an individual sales person to uncouple from the strictures of operating policies and procedures followed by the rest of the sales personnel or the rest of the organization. As you work to change the type of coupling at one boundary be conscious of the effects across other boundaries.

All complex systems include boundaries at many levels. You will find coupling relationships tend to scale across an organization. For example, if the departments within an organization are uncoupled from each other, you will probably find the same organization uncouples from its customer, individuals uncouple from each other and teams tend to work in isolation. For more information about the scaling phenomenon in complex systems, refer to Chapter 4, which talks about scaling as a fractal structure.

Coupling satisfies emotional requirements of individuals within an organization as well as institutional goals. You will find some persons are most comfortable working in uncoupled isolation from peers. Others relish the interdependency of a loose couple. Others work most comfortably in the power relationships of tight couples. Consider the emotional and psychological impacts of changing coupling relationships within an organization.

One-way behavior will not establish a couple or change its nature. Both sides of the boundary must negotiate an effective coupling relationship. If you find yourself unsatisfied with a coupling relationship, and the other part of the system is not willing to change, work on establishing basic transforming feedback loops across the boundary before you seek to alter the more complex coupling relationship.

Over time you will begin to discern the special characteristics of the coupling relationships that make your organization productive. Consider the impact and use your communication network to watch for changes in coupling relationships as your organization adapts to meet changes in the economic, technical or political environment.

Coupling: An Example

JET, a large, midwestern financial management company, grew to great success in the 1970s. Its products included life insurance, certificates of deposit (CDs) and a variety of equity funds. Each of the three product families was managed as an independent business. They shared office space, a consumer-oriented sales force of independent agents and a single Board of Directors. In all other respects the Insurance, CD and Funds organizations functioned independently. The home office, which provided different services to support each of the different products, was perceived as a remote appendage. Communication between

160

the home office and the agents consisted of infrequent and punitive written documents. The home office sent procedural manuals and policy statements into the field. The agents sent hand-written orders to the home office for processing. Otherwise, communication was kept to a minimum. No love was lost between agents and employees of the home office.

This arrangement worked well during the 1970s. Sales volumes were relatively small, customers tended to purchase only one type of product, and sales agents focused on selling the product they knew the best or felt most comfortable with. In the early 80s, however, several aspects of the business changed.

- The organization was purchased by an international conglomerate that expected radical decreases in operating expenses.
- Customers grew more sophisticated about their investment decisions and began purchasing a combination of products from JET.
- Computer technology became available to allow faster, more reliable on-line communication between the home office and the agents.

The uncoupled nature of the organization worked against success in each of these arenas. The parent company saw that the arch independence of the product companies resulted in redundancy and unnecessary overhead expenses. Customers wanted to be able to receive consolidated financial statements reflecting their investments across the company's products. Agents struggled to learn the variety of procedures and policies for all of the product companies and complained they felt they were working for three companies instead of one. Differences in internal procedures and ways data were handled made it quite difficult to create integrated computer systems to link the three product companies together.

The Board of Directors decided the best way to solve the problem would be to force the product companies to work together. They chose to implement this strategy by developing an integrated computer system to handle sales, customer information and accounting data for the company as a whole. The Board believed this system would force the product companies to work together. In short, the computer would form the tight couple to provide the integrated business environment for the future.

JET invested seven years and many millions of dollars to create a custom computer system to meet the diverse needs of the three product companies. Teams spent endless hours discussing the special needs of each product and looking for compromise solutions to allow the same system to meet all of the needs of all the groups. In short, they were looking for a way to use technology to create a tight couple among the product companies. The result was a system that supported the lowest common denominators but failed to meet the unique needs of each group. Each product company was forced to create additional modules to customize the system to meet their needs. Customer communication across product lines was improved, but the product companies themselves remained uncoupled.

The next major impetus for change came in the late '80s, when new markets and new players heated up competition in the financial industry. JET decided individual customer service would provide the competitive edge they needed in the market. Agents would be expected to provide integrated financial management services to all clients. From the perspective of agents and clients, JET's products and services had to present a seamless whole. Before JET could provide this level of integration, the product companies would have to be coupled in some way.

Again, the Board of Directors intervened. This time, they turned to organizational development technology for a solution to their problem. Following the tenets of sociotechnical systems, JET instituted a total redesign of their customer services processing units. They created teams who would be responsible to provide service for all JET products within a geographical region. All employees would have to understand the special characteristics of every JET product. The equity specialists had to learn the details of CD product processing. The CD gurus had to sell and service equities and insurance products, and the experts in insurance were expected to become experts in the equity market and the CD environment. This, too, was an effort to create tight couples where none existed before. JET invested enormous resources in cross training and absorbed the cost of rework when naive personnel tackled sophisticated product-specific problems.

Within a year, the "everyone must know everything" philosophy fell into disrepute. But the complex interdependencies and training processes generated a solution by establishing loose couples among the product companies. The geographical teams would handle customer information questions and simple product questions. Complex product questions would be routed to a group of experts for resolution. Insurance experts would handle insurance questions. CD experts would resolve CD issues. Equities specialists would respond to questions about mutual funds. This combination of tight, loose and uncouples resulted in excellent and efficient customer service and support to agents in the field.

Even the uncouple between the agents and the home office slowly shifted to a loose couple. The geographical teams got to know the agents who worked in their areas, so informal associations and friendships developed. Information flowed from the home office to the field and back through these informal communication networks, rather than the formal management memos and agent complaint forms of the past.

Through the applications of technology and sociotechnical systems, JET was able to resolve the coupling problems that had plagued it for over ten years. If, however, they had recognized the characteristics of their complex system and dealt directly with issues of boundary conditions and coupling, the change process could have been significantly more timely and less expensive for all concerned.

When you understand and respond to organizational influence in the form of coupling relationships among organizational parts, you will be better able to anticipate and respond to unexpected influences in the environment. Much as the pilot of a sailing ship recognizes the signs of impending change or smooth sailing, you, too, can read the environment and respond wisely to the shifting winds of change.

Epilogue
Chaos at Work

Epilogue
Chaos at Work

Frank Dubinskas entered my life as significant people often do, by accident. Friends and colleagues had mentioned a new professor at Hamline University, an anthropologist who was interested in chaos. A promising potential employee, Anna Hargreaves, mentioned that her husband was applying chaos to organizational questions about the integration of technology and manufacturing at Apple Computer. At a meeting of a professional organization, I saw Anna and met Frank briefly. In the midst of the bustle and cross-conversations, we arranged to have lunch.

That extended luncheon conversation was the first and last opportunity I had to talk with Frank about his work. He was applying the metaphor of chaos science to understand the twists and turns of Apple's Erickson Project (the Erickson Project is discussed here courtesy of Apple Computer, Inc.). Frank had spent about 18 months working part-time, on-site at Apple to observe the progress of a massive manufacturing construction project. In his early letters to Apple he says, "My mission is understanding and improving complex project management, including the relation between engineering designers, the work of operators who will eventually use the system, and the tools they will use to do it." The data he collected was rich with chaos applications in real-world settings. Frank's interpretation of the data was intriguing, well-researched and profound.

When I left the restaurant and stimulating conversation, I was convinced Frank would be a perfect mentor in anthropological research techniques and a wonderful colleague in my search

167

to apply the tenets of complex systems theory to everyday management problems. I knew Frank was ill, but his enthusiasm for the subject and open curiosity about ideas masked his pain and weakness. Before we were able to expand our work together, Frank's physical strength failed him, and he died in October of 1993.

Frank's wife Anna, now a colleague and friend, sorted through and catalogued Frank's papers. She asked me to help edit some of his later works for publication and to work with her to present a memorial retrospective of Frank's work at a management conference. During this process, Anna expressed a desire for Frank's data to be used to further the work he loved so well. At the same time, I was looking for an extended case study that would tie together the concepts of this book and to help make the ideas and techniques real for readers. This chapter is designed to meet Anna's goals and my own. It is dedicated to the lasting memory of a man who embraced the beauty and power of complexity in all aspects of life—Frank Dubinskas, who is deeply appreciated and sorely missed.

The Erickson Story

In 1987, Apple Computer began to reassess its manufacturing strategies. Initially, in the spirit of state-of-the-art focused manufacturing, they had designed manufacturing environments that were optimized for a specific product group. With the broadening of their product line, they realized the need to design flexible manufacturing environments. The same line would support manufacture and testing of multiple products with minimum downtime between products. Looking for a new strategy, Apple considered extensive automation and computer-aided manufacturing. Management and technical discussions covered the available options for improving both the efficiency and flexibility of the manufacturing processes. By late 1989, competition from other manufacturing facilities forced the plant in Fremont, California to settle on a plan quickly. In December of 1989, the Erickson core team was formed and work began on the project that would transform the Fremont manufacturing process and ultimately transform all of Apple's manufacturing processes.

The functional specification for the new line was complete by the end of January, and a Japanese design partner had submitted an acceptable bid for constructing the line. The design was approved in late April. Acceptance testing was completed in Japan by mid-June. The whole line was disassembled in Japan, shipped to Fremont, reassembled and tested. Production started by mid-September. The new product was launched in October of 1990.

The product of this enormous project was a computer-integrated machine that was over a football-field in length and three stories tall. This machine performed assembly, burn-in and final testing of new Macintosh™ computers. Two control systems drove the assembly. First, a machine-controls architecture operated the equipment. Second, an autonomous Mac-based factory management system handled data about the products as they moved through the line. All of this massive project was completed within schedule and budget, without a single engineering change request.

The project was not without its complications, but on the whole, the project team met all of its explicit objectives. From this perspective, the project was a definite success.

Less than a year after the line became operational, Apple management decided to close most of the Fremont site. Line 1 was scrapped because it was too big to move to another building, and it had no value to potential buyers. From the perspective of long-range contribution to manufacturing efficiencies at Apple, the Erickson Project was a total failure.

This story incorporates two amazing sets of events. First, how was the team able to construct the line with such reliability in such a short time? Second, how could this remarkable feat end up on the scrap heap? Traditional, Newtonian explanations are insufficient to explain either of these situations, but the principles of complexity shed light on the dynamics that drove both of them. In the following sections, we will apply each of the principles of complex systems from preceding chapters to help you understand how these two, seemingly contradictory situations emerged from the same complex environment.

Butterfly Effects

The butterfly and its effects were alive and well in Apple throughout the Erickson Project. In a project of this magnitude, small causes and their enormous effects are common. We will focus on ones that affected the product strategies that initiated the project, ones that affected the specifications for Line 1, and those that contributed to the decision to scrap the line so soon after its successful introduction.

Product and manufacturing strategies. Manufacturing strategy at Apple had historically been tailored to new product introductions. Product strategy would be defined one to three years in advance. Based on the anticipated product line, Apple would evaluate and make plans to upgrade facilities to support manufacturing requirements. One of the primary problems with this well-reasoned approach was determining the new product strategy. Any small problem in the accuracy or stability of new product planning generated far-reaching effects in the manufacturing processes and investment in facilities. Any shift in demand forecasting had enormous effects on production volume requirements and turn-around times.

170

Planning from 1987, which drove the Erickson Project, indicated a need for high-end Mac products with an infinite number of configurations and many small items, such as external hard drives. Manufacturing would have to support a flexible product line with the ability to schedule builds on the basis of large-scale business customers' requirements. These logical, and relatively minor, shifts in product strategy were strongly reflected in the engineering specifications for Line 1. The marketing butterfly flapped its wings and the manufacturing hurricane ensued.

Line 1 specifications. The technical specifications for Line 1 were revised many times before actual construction began. After the original Erickson Project proposal was complete, budget constraints were imposed. Some cuts could be absorbed without affecting the basic specifications, but at one point, the loss of an additional $250,000 forced fundamental changes in the specifications that had the butterfly's ripple effect throughout the project.

Mid-project, a problem arose that had not been adequately considered in the original specifications. The problem dealt with defective parts that might be identified at or near the beginning of the line. The core team at Apple wanted to redesign detailed control mechanisms to track these defective parts. If the project had taken this turn, the missing specification would have required the redesign of significant parts of the system. The Japanese engineers, on the other hand, came up with a solution to minimize effects on the overall design. The Apple core team disagreed vigorously. Discussions around the issue continued for some time and resulted in a workable solution, but one that did not satisfy several members of the core team.

This butterfly incident generated emotional and personal effects in addition to the technical ones. Attitudinal shifts and frustrations are clear from communications at the time, but we have no data about the long-term impact of these emotional

171

effects. We can imagine, however, that they might have caused problems in communication long after the technical problem had been solved.

Decision to scrap Line 1. Many variables affected Apple's decision to scrap Line 1, but one is clearly based on a butterfly effect. In October of 1989, just as Erickson was getting off the ground, a 6.9 earthquake struck the East Bay. Apple planners were reminded that their whole U.S. manufacturing capability was only a few miles from the Hayward fault. Apple decided to open a second manufacturing site in Colorado Springs, which had lots of room for expansion, was cheaper to operate and was not known to be earthquake-prone. Since Line 1 could not make the move to Colorado, it had to be scrapped.

Recommendations. None of these butterflies could have been predicted or avoided. Apple had to base its manufacturing decisions on long-range product planning, even though they anticipated variations over time. Budget constraints do affect the technical characteristics of complex products. It is not cost effective to spend inordinate amounts of time in a specification stage to ensure all contingencies have been covered before work begins. In a complex system, no amount of time will be enough to predict all possible outcomes, anyway. Tough technical decisions and the conversations preceding them will always generate emotional reactions from dedicated personnel. The chaotic nature of earthquakes means they can be anticipated but not predicted completely.

Small changes in a project environment sometimes have profound effects on persons and products involved. Apple management and the Erickson team managed these butterflies effectively. They adapted and adjusted corporate strategies and project plans to respond to the unforeseen effects of butterflies.

Boundaries

A variety of boundaries were established in the Erickson Project structure, and others emerged in the course of the project. In many cases, the conditions around these boundaries provided energy and insight. In other cases they distorted information and resulted in misunderstanding or delay. We will focus on three boundaries that are most evident in the data—those between individuals, between humans and technology on the line, and those between Apple and their Japanese design vendor.

Individuals. The Apple core team was intentionally kept small to minimize the effects of individual boundaries. The team included both men and women, persons of various racial and cultural backgrounds, engineers from a variety of disciplines and representatives of different organizational groups. With a clear goal and limited time, the team overcame these boundaries quickly to form an efficient working group. Throughout the project, however, the existence of the boundaries was quite evident. Organizational groups raised issues related specifically to their institutional roles. Each engineer sought to optimize his or her component of the design and showed resistance when compromises had to be made. Communication styles and culture-based behaviors shaped the focus and progress of conversations about technical and project-related issues.

Electronic communications were used extensively throughout the project. Electronic mail and voice mail formed the basis for crossing communication boundaries. References were made to the inability to reach team members by voice mail, and individual members of the core team were not allowed direct e-mail access to all members of the Japanese design team. One representative even said it was illegal to exchange data internationally. Regular team meetings and other forms of formal communi-

173

cation filled in the gaps when electronic means were not sufficient to meet specific project requirements.

Humans and technology. The boundary between humans and their technology was a major focus for Frank in his studies. In an early communique, he states,

> I am interested in matching appropriate human systems integration to technical/computer systems integration—that is, linking people in concert with an integrated manufacturing system. This interest has several more specific foci:
>
> - Organization and integration of project teams which include different functional expertise,
> - Building technical systems which serve this integration by facilitating collaborative action, information and knowledge sharing.
>
> My sense of the field is that our knowledge of integrating groups (rather than just individuals) with manufacturing systems is far weaker than our ability to integrate the technical systems themselves.

Throughout the project, Frank identifies the need to use the technology to broaden the participation of operators on the line. The Erickson team focused on technology as a way to collect data for analysis by engineers. Frank asked questions about how the technology might be used to provide information for capture, trouble-shooting and problem-solving by operators.

Another aspect of human/technology boundary fits into the design method employed for the project. After the line was reconstructed in Fremont, testing involved operators. Their input identified important options for improving the mechanism based on the needs of those who would do the on-going assembly of products. The fruitful outcome of such testing late in the project led project team members to recognize their input might have been helpful earlier in the project. Engaging the assemblers—real users—at the earliest possible stages would have created "design for implementability," with people in mind. Such user-centered design is now widely recognized as an effective technique to minimize the boundary between human beings and the technology they use to do their work.

Apple and the design vendor. An unavoidable boundary exists between any vendor and customer. In the Erickson Project, that boundary was exacerbated by geographical and cultural differences. At many points, team members expressed excitement about the conditions around the boundary. For example, one team member described the first technical negotiation meeting in the following words.

> . . .There was a lot of 'unexpected energy' around the meeting. Not exactly surprises, but the event was like a steeplechase: The rider knows he's coming up to a barrier and knows he's going to leap and knows he can... but there's still a charge of energy and excitement involved in going over! Erickson may have nicked the bar a little bit, but they got over!

Initially, the core team perceived the boundary with the vendor as clear and distinct.

The original specification was planned to have several milestones, referral of in-process partial completions to Apple for approval and a final acceptance and approval of the system in Japan. This reflects a common American approach to vendor relationships in which the customer defines the project carefully up front; limits uncertainty and variability as early as possible to limit surprises; and then holds the vendor to the contractual agreement. With such a boundary, management becomes a process of variance control and legal interpretation.

This contrasts with a more typically Japanese-to-Japanese style of customer-supplier relationship, which includes a higher tolerance for design uncertainty farther into a project. Alternative

175

ideas can be introduced later into the project. They assume changes are inevitable, so they anticipate and prepare for change rather than trying so hard to legislate it out of existence with detailed specifications and contracts. This anticipated flexibility provides a platform for negotiation and for a discovery process, which are envisioned as leading to joint learning. The relationship requires a great deal of trust and commitment between the partners, since much proprietary information is exchanged, and neither can easily escape the relationship. The strategic aim is to build close links for ongoing mutual competitive advantage and to learn from each other to build internal expertise in areas where one is not fully competent.

Over the course of the project, the Japanese definition of the relationship won out because schedule and budget constraints took precedence over some of the technical legalities. Some members of the Apple team resisted or resented the flexible relationship. They perceived the loss of control over the product and project process as a necessary, but unfortunate, compromise.

The Apple core team attended cultural training and orientation sessions early in the project to avoid some of the destructive conditions at the cultural boundary. Still, some problems with cultural differences emerge from the data. The Japanese engineers were willing to use the operators to trouble-shoot non-standard problems on the line, while the Apple team looked for fully automated solutions. Frank observed and team members commented on the reluctance of the Japanese to say "no" directly during negotiations and their tendency to indicate understanding when they really did not understand. The vendor chose to construct and test the line first in Japan, within the range of their own environment and personnel. Then they broke it down and transported it to Fremont for re-assembly and on-site testing. The translator between English and Japanese speakers was an engineer who happened to be bilingual. Some of the Apple team members expressed concern about the accuracy and completeness of the translations. Boundaries existed within the vendor's organization between their Japanese and U.S. offices. Non-technical personnel who were part of the U.S. division formed the primary communications boundary between Apple and the Japanese engineers. This approach to managing the boundary introduced some interesting and frustrating delays in the project.

Recommendations. All of these boundaries among individuals, between the technology and human beings and between customer and vendor affected the flow and efficiency of the project. Through the lens of complexity science, we might recommend the following actions to minimize the disruptive boundary conditions and maximize the constructive ones.

- Explicitly investigate the myriad boundaries between individuals on the team and between the team and other working groups related to the project.
- Provide consistent e-mail and voice mail access between individual members of the teams.
- Integrate the operators into the system specifications and design by creating an interface to provide them with information during manufacturing and including them in early testing of prototypes of the system.
- Acknowledge the need for a flexible relationship between vendor and customer early in the process to optimize mutual learning and reduce time and energy focused on legalistic framing and interpretation of specifications.
- Put a member of the Apple core team on site in Japan to observe and communicate with the vendor's engineers.
- Include a professional, objective translator to serve as a support resource whenever the teams meet together.
- Set up primary technical communications between engineers who are involved in the project directly, not between management personnel.

177

Transforming Feedback Loops

A transforming feedback loop describes the communication across a boundary which allows both sides to adapt and adjust to the needs of the other side. Efficient transforming feedback supports all of the work of a project team from inception to completion. Erickson was no exception. We will consider a small selection of feedback loops and their effect on the Erickson Project schedule, budget and personnel.

Contract negotiation. Two sets of feedback loops were critical during this stage of the project: those between management and team members within Apple and those between Apple and the Japanese vendor.

We have already discussed the marketing and management considerations that preceded the initiation of the Erickson Project. Distortions and timing of those feedback loops caused the delay from early discussions of flexible and automated manufacturing until the project start in December of 1989.

Even after the request for proposals was distributed and the vendor selected, additional delays were caused by administrative decision-making procedures. Commitments were made and remade, affirmed and reaffirmed before the technical work on the project could begin. These delays affected the overall project schedule and certainly affected the motivation, commitment and confidence of the members of the core team.

After the contract was finalized, the discussion and clarification of specifications involved some long and unreliable feedback loops. At first, the Erickson core dealt with one American-born vendor engineer. He worked through his U.S. supervisor, who transmitted communications or specifications to Japan. The specs were translated into Japanese, and engineers there began to work on them. This four-step process was required even before concrete work began. Errors and distortions accumulated during the process. For example, a month was lost when the Japanese engineers did not examine the specs as closely as Apple had expected. Subsequent misinterpretation of the specifications required rework during the early design phases of the project.

Project progress and status reporting. Even after the project began, communications were slow and somewhat cumbersome. It would take three to four days for a question to be asked and transmitted to Japan. The response required the same amount of transmission time. Management was not worried about the turnaround time because they felt the work of the teams was sufficiently independent for the delays to have minimal impact. As the project continued, however, the two groups would have been able to work more cooperatively if the feedback loops had been shorter and more reliable.

At one point, in fact, a planned trip to Japan was postponed when the core team discovered that the vendor was not as far along as was expected. During a meeting in Fremont several questions and issues had arisen that had not been sufficiently addressed. The engineers requested an opportunity to work on the issues with their team before the Apple visit. They chose to lengthen the feedback loop to allow for more time to prepare adequate responses.

The on-site visits of the Apple core team in Japan seemed to provide excellent opportunities for building transformation into the feedback loops. In spite of language and cultural differences, the team always expressed satisfaction after having opportunities to work with face-to-face, short feedback loops. Not only did they provide personal reinforcement, they also allowed many technical issues to be aired and resolved quickly.

Product testing. Testing, in any development method, establishes formal and informative transforming feedback loops.

In the Erickson Project, testing feedback loops generated trans-formations in specifications as well as in project process assumptions.

The permeable boundary between Apple and the vendor supported a variety of learning transformations for both the Japanese and the North Americans. These late changes allowed the product and the process to adjust for newly developing knowledge from internal sources. As the groups worked together on the project, they discovered new things as parts of the product were created. In this way, Erickson was able to work within a system of feedback loops to engender transformative learning within the constraints of the project.

In particular, operators who tested the line in Fremont asked for more visual, tactile and auditory cues to support the assembly process. The controls specifications and design had not incorporated extensive sensory output because they are not critical to design for robotic assembly. Robots do not look for a position, feel a part click into place, or listen for the smooth release of a component. When humans completed the task, however, they depended heavily on their sensory feedback mechanisms to complete the complex physical tasks. The testing provided engineering feedback on the need for assembly feedback. Not only was this information used to transform the assembly process, it helped project managers recognize the need to include the operators in early specification and design stages of the project.

Organizational learning. Early in the Erickson Project it was clear the team members considered themselves to be creating more than a single manufacturing line. They were engineering a new approach to manufacturing at Apple. They anticipated that the technological and management innovations they discovered in the course of the project would transform Apple's manufacturing strategy. They expected to establish transforming feedback loops to affect the entire corporation.

180

Apple supports a variety of cross-functional communication teams to support this kind of transforming feedback. The Manufacturing Council meets quarterly and includes top manufacturing managers from around the world. They discuss manufacturing strategies and tactics. This is one of the arenas in which the Erickson Project was conceived. The Site Engineering Review meetings were held periodically at Apple. All supervisors and selected engineers met to ensure diffusion of information and encourage greater engagement among engineers and institutional components of the site.

The Erickson Project went beyond these regularly-scheduled feedback opportunities for organizational learning by working with Frank Dubinskas. Frank not only collected data about the project, he also evaluated and reported on ways the project and product development processes might be improved.

Recommendations. During contract negotiations, project work, testing and organizational learning efforts, Apple and the Erickson team sought to establish and use effective transforming feedback loops across personal and organizational boundaries. In addition to the efforts described above, the following suggestions might increase the power of transformation through feedback.

- Establish on-going feedback loops between product strategy development and manufacturing. The initial information about future products got Erickson going, but later changes were not reflected in project adaptation until after the project was complete and the line was installed.
- Shorten the feedback loop for decision making. Precious weeks and months were consumed with management and administrative decisions early in the project, limiting the amount of time available for technical design, development and testing.
- Establish shorter and less distorted communications among technical team members, or plan for more frequent on-site meetings.
- Expect innovations to come as changes late in the project. Establish a procedure to collect and evaluate requested changes so the valid innovations can be integrated into the current or future projects.

- Disseminate information about the celebrations and mistakes encountered during projects. Both provide effective learning opportunities for organizations beyond a current project. A formal post implementation review to summarize and evaluate the project process and the resulting product characteristics would help capture relevant information for sharing across the organization.
- Keep the project team conscious of the need to learn from successes and failures in the course of the project.
- Incorporate prototyping and early user involvement to ensure transformation in line with feedback from real users.

Fractals

In real life projects, it is difficult to see the beautiful fractal patterns because you are too close to them. Sometimes one must step away from an individual project to observe coherent fractal structures. Members of the Erickson team noticed examples of scaling and self-similarity, but executive spokespersons at Apple were the ones who described the pattern emerging from the whole.

Deborah A. Coleman, Vice President, Finance of Apple Computer, Inc., described the coherent Apple vision at the Conference on Technology in Business Education on December 6, 1988.

> We've undergone an enormous amount of change and growth, but through it all we've held on to the fundamental vision upon which the company was founded. That vision was—and still is—to make the power of information technology accessible to the individual, to create tools for the mind and to change the world, one person at a time, with our products and technology.

This describes the fractal generation of an organizational entity—a simple, but replicable concept to be used in a variety of contexts. This dynamic process generates an organization that exhibits scaling and self-similarity. It demonstrates coherence in the midst of diversity.

In the same presentation, Ms. Coleman described one of Apple's underlying beliefs, "creatively productive individuals contribute to improved organizational performance." Here is a wonderful example of scaling at work. Productivity at the individual level generates the same behavior at the organizational level.

In February of 1989, Morris Tadalasky, Apple's Vice President, Customer Services and Information Technology expanded this concept of scaling to global proportions.

> Conceptually, Apple's goal in its product development is perfectly in line with the motivation for a unified Europe. Just as we strive for a barrier-free access to information, and a consistent user-centered experience, the European Community envisions a free-trade area for the movement of goods, people and services from country to country without tariffs, duties, or restrictive regulations.

Mr. Tadalasky recognizes the importance of boundaries at various scales: individual to national.

These broad brush strokes of corporate coherence show themselves also within the project behavior of the Erickson team. These scaled reflections are not always beautiful and productive.

In one of his review presentations, Frank noted,

> It is especially difficult for organizations such as Apple Fremont, which are dominated by an engineering or professional culture, to value knowledge building for continuous improvement at the operator level.

He recognized how the engineer introduced complicated boundaries by looking for self-similarity in all problem solvers.

Product and manufacturing self-similarity at Apple has been seen as a competitive advantage, which contributed to the importance of flexible manufacturing strategy of Erickson. "A user can take a program that was developed on a 1984 Macintosh™, insert it into the most recent product and perform the same task, only at a faster pace and with greater ease." (Tadalasky, 1989)

The Japanese American engineer was selected to coordinate the vendor's communication with Apple. To what extent did the racial self-similarity to the core team contribute to this decision?

In the course of the project, members of the core team overcame their individual differences and emerged as a coherent, unified team. Focused on the single goal of completing Line 1, they worked as a monolith in some respects. This self-similar behavior contributed to the myopia that kept them ignorant of the global and corporate changes that would make the line irrelevant soon after its completion.

Recommendations. The fractal nature of a complex system brings with it the beauty of both self-similarity and diversity. Patterns that can be discerned at macroscopic scales will be difficult to see immediately at microscopic ones. Self-similarity that becomes obvious at microscopic scales tends to close perceptions to the macroscopic levels of diversity and variation. The following recommendations will help individuals, teams and corporations use their fractal natures to improve the ability to adapt.

- Periodically restate project objectives in terms of the objectives of the corporation.
- Do not allow self-similarity to mask critical instances of diversity.
- Use possibilities for self-similarity to cross obvious lines of distinction among classes or groups of employees. For example, recognize that operators on the line have a similar concern for quality and ability to solve problems as engineers who design or plan maintenance for the assembly line.

Attractors

Many different attractors affected the Erickson Project and the technical features of Line 1. Each had a part in shaping the events that occurred before, during and after the project itself. Examples of each of the three types of attractors are described below.

Point attractors. The most obvious and powerful point attractor for the Erickson Project was its schedule. The end date was set early in the project. It became a focal point for all of the project team and their activities. Such a clear and distinct point attractor has its positive effects. Tight schedules lead to strong commitment from all of the members of the Erickson team. They provided the personal and organizational motivation for effective work to realize on-time delivery of the product. A restrictive schedule, however, introduced complications into the project process.

- Technical decisions were made in the interest of time that, in some team members' views, compromised technical quality of the system.
- Reduced time for communication and coordination restricted the teams' ability to incorporate insights from other groups within Apple. Insularity is one potential effect of such clear and distinct point attractors.
- Opportunities for creative negotiation and organizational learning were minimized or ignored completely.

Periodic attractors. Corporate policies and procedures drove periodic attractors throughout the project. The weekly team meetings set the pace and structured the communications of the core team. The annual budget cycle drove the project planning and execution procedures. Regular reports to management influenced the flow of events internal to the project. And, on a larger scale, the product planning and development cycles generated interest in the new Line 1, and, ultimately, brought about its destruction. All of these contributed to the ebb and flow of the project life and the team members' activities.

Strange attractors. Strange attractors are much more difficult to spot than point or periodic ones. Behavior driven by a strange attractor appears to be random, but over time, patterns of behavior emerge. Individual data items are insufficient to identify the presence of a strange attractor. Instead, an accumulation of many data points and their complex relationships over time give evidence of a strange attractor at work.

As I studied Frank's data about the Erickson Project, underlying patterns emerged. Early in the process, each e-mail note and meeting agenda was unique and contributed specific, quantifiable information to my growing understanding of the project. At some point, however, a pattern of the whole emerged. After that point, new data served to reinforce the pattern or to enrich it with new detail or surprising turns. Still, the pattern is there.

I have had the same experience in projects I have worked on or observed closely. Initially, events are scattered and not coherent, but over time they coalesce into clear, identifiable patterns. Each project has its own, unique pattern and a kind of inevitability built into events as they spread themselves over time. When I describe projects to others, I present details as I remember them, but my personal understanding of the project is stored away as the overall, emergent pattern. This, I believe, is the strange attractor. I expect everyone involved in Erickson carries a gestalt memory that incorporates all of the forces and patterns they perceived as the project progressed. Are these individuals' patterns identical? I expect not. Do the patterns incorporate some fundamental level of self-similarity? I expect so.

Recommendations. Any group activity within a corporate setting must respond to a variety of attractors. Undue focus on any one of them tends to throw the activity into unproductive patterns of behavior. Based on this perception, the Erickson Project was too driven by schedule. This point attractor excluded opportunities for the project team and the project itself to adapt to changes outside of the project boundaries.

Self-Organization

Self-organization is the tendency for a far-from-equilibrium complex system to generate new structures spontaneously. The Erickson Project included many instances of self-organization, we will focus on only four here.

Formation of the team. Within a span of less than two months, the Erickson team had "consolidated itself" as separate from a larger, more diffused team, which had existed earlier. This consolidated team included four members, one from each of the primary disciplines (Computer-Integrated Manufacturing, Maintenance Engineering, Facilities and Systems Integration). Within another month, the team had defined the vision and goals for Erickson, and they were ready to document the system specifications. After almost three years of technical and management discussion, this group was able to structure the massive project scope within a matter of weeks.

Technical specifications. Early statements of system specifications are uncertain, at best. Their details evolve and emerge over time as a project progresses. Conversations (sometimes heated) between the vendor and the Apple team stretched original specifications and implementation plans past comfortable equilibrium. Out of these complex and heated discussions, a reasonable set of system specifications arose.

Relations between the vendor and Apple. Disagreements about control architecture and tools arose early in the project. As these issues were resolved a new relationship emerged between the two working teams. While a commitment to joint learning was not central to the original Apple plan, it evolved spontaneously out of the ongoing work.

Scrapping the line. The ultimate and most dramatic example of self-organization in the Erickson Project is the scrapping of Line 1. As the project progressed, changes in the rest of the system continued. The U.S. economy, the geological stability of the area around Fremont and the management personnel and structures of the rest of Apple all moved farther from equilibrium. The project team, focusing on their technical specifications and their end date, did not notice as the other parts of their larger system moved far from equilibrium. The team was shocked when the self-organization occurred, and manufacturing shifted from Fremont to Colorado Springs. The Line was sent to the scrap heap. The team's one-time success shifted to failure.

Recommendations. To make the most out of opportunities for self-organization, the Erickson team might have planned and managed for self-organization. How? Use formal and informal communication networks to keep the subsystems connected. When self-organization does happen, all parts of the system will be expecting the shift and will have an opportunity to prepare.

Coupling

Patterns of transforming feedback form semi-permanent relationships between parts of the system. Because the Erickson Project incorporated so many different groups and such rich feedback loops, couples of various kinds emerged in the course of the project. The sections below describe examples of each of the three types of coupling relationships found in complex systems.

Tight couples. A tight couple indicates a relationship in which each of the parts of the system exerts strong influence over the other. Examples from the Erickson Project are described below.

- Presentations by senior management indicate they perceived a tight couple between Apple and national economic interests world-wide.
- At the beginning of the project, Apple assumed they would have a tight coupling relationship with the vendor by defining and controlling specifications.
- Management concern over schedule and budget established a tight couple between the project team and the administrative and control functions of management.
- Working closely together, members of the core team developed tight couples between pairs of individuals. These relationships, however, shifted over time as issues arose and allies were redefined.
- The Apple core team sought a design for Line 1 that would be tightly controlled by complex, thoroughly automated systems.
- The nature of the technology and the interdependence of the controls required a system in which parts of the Line 1 were tightly coupled to each other.

Loose couples. A loose couple indicates the two parts of the system influence each other, but both are open to other influences as well. Examples from Erickson are listed below.

- The engineering vendor promoted a decentralized architecture which could be implemented in pieces. Their implementation procedures included manual tasks in which the operators on the line intervened in the workings of the line whenever necessary.
- Installation of the Line 1 at Fremont had to be completed without interrupting normal manufacturing processes. Coordination between on-going production and installation and testing of the new line was critical. This relationship was a loose couple.

- The relationship between good team leaders and their teams indicates the need for loose, but reliable coupling.
- Periodic meetings between the Apple core team and the vendor resulted in loose coupling over time. (This is an interesting development when the original expectation of Apple was that they would have tight coupling regarding technical issues and uncoupled relationships with regard to communication and project process.)

Uncouples. In an uncoupled relationship, neither part of the system affects the other. Several examples of uncouples are described in the Erickson Project data.

- The gulf between engineering groups and line operators is inadvertently reinforced by Apple's "engineering excellence" culture.
- One-way links between business and manufacturing planning result in lack of communication over time. Business planning started the Erickson Project, but it also intervened two years later to shut down Line 1.
- Some of Apple's technical development groups were "blind to the rest" of the Fremont groups. These uncouples led to distrust, lack of informed decisions, occasional rework and high levels of frustration.
- During the installation and testing of Line 1 at Fremont, a strongly worded message went out to all Fremont employees, "If you are not directly involved in the installation, stay away to let others do their work. If you don't belong, a Production Supervisor will ask you to leave." This procedure sought to uncouple the Line 1 production from the other activities at Fremont until the line reached some steady-state production status.
- During the project, Erickson team members and management seemed unaware of the changes in management and marketing strategies that ultimately caused Apple to scrap the line.

Recommendations. Advice here is simple: Try to maintain loose couples in all directions. They are easiest to maintain, provide best on-going information and support smooth adaptation when it is necessary. Shift to a tight or an uncouple when there is a clear and distinct reason for more control or less interdependence.

Conclusion

Frank Dubinskas and the Erickson team worked together and met a demanding collection of objectives in the midst of a turbulent and chaotic environment. They executed their tasks effectively within a bounded region. They focused on their project goal, treating it as a point attractor. Within this context, their work lead to glorious success.

From a different vantage point, within a larger context, the value of their labors was short-lived. In retrospect, using the lens of complexity science, the weirdness over which they triumphed is understandable, even if it could not have been predicted. Complex interdependencies beyond their control literally destroyed the fruits of their labors. In time, other interdependencies, equally complex, may emerge to transform their learning and the products of their work into other, unimagined creations.

In the same way, you labor in a world of complexity. You will apply all of your resources to meet predetermined goals. Sometimes you will be delighted by unexpected events. You will be disappointed by others. Your expectations will adjust to match the world as you understand it, just as the world turns in a different direction. Complex interrelationships in your personal and organizational systems will lift you up, and they will turn you over. A pattern of unpredictability is the only thing you can predict. To live productively in this chaotic world, you must collect information, take decisive action, and collect information again in a transforming cycle of learning and growth. I hope that *Coping with Chaos* has given you insights and tools that will help you learn and grow more productively in your own unique and complex environment. How will you use the insights of complexity and these seven simple tools to meet the challenges of leadership in your corner of the unpredictable universe?

Bibliography

Abraham, Ralph. (1994). *Chaos, Gaia, Eros*. San Franisco: Harper.

Bak, P. (1996). (1996). *How Nature Works: The Science of Self-Organized Criticality*. New York: Springer-Verlag New York, Inc.

Briggs, J. and Peat, F. D. (1989). *Turbulent Mirror: An Illustrated Guide to Chaos Theory and the Science of Wholeness*. New York: Harper Row.

Briggs, J. and Peat, F. D. (1984). *Looking Glass Universe*. New York: Simon & Schuster.

Casti, John. (1995). *Complexification*. New York: Harper Perennial.

Davics, P. (1983). *God and the New Physics*. New York: Simon and Schuster.

Douglas, M. (1986). *How Institutions Think*. Syracuse, New York: Syracuse University Press.

Gell-Mann, Murray. (1994). *The Quark and the Jaguar*. New York: W. H. Freeman and Company.

Gleick, J. (1988). *Chaos: Making a New Science*. New York: Viking Penguin.

Goldstein, J. (1994). *The Unshackled Organization*. New York: Productivity Press.

Guastello, S. (1995). *Chaos, Catastrophe, and Human Affairs*. Mahwah, New Jersey: Lawrence Erlbaum Associates, Publishers.

Hall, Nina (ed.) (1993). *Exploring Chaos: A Guide to the New Science of Disorder*. New York: W.W. Norton and Company.

Hayles, N. K. (Ed.). (1991). *Chaos and Order: Complex Dynamics in Literature and Science*. Chicago: The University of Chicago Press.

Kaplan D. and Glass L. (1995). *Understanding Nonlinear Dynamics*. New York: Springer-Verlag.

Kauffman, Stuart. (1995). *At Home in the Universe*. New York: Oxford University Press.

Kellert, Stephen. (1993). *In the Wake of Chaos*. Chicago: The University of Chicago Press.

Kelly, Kevin. (1994). *Out of Control*. New York: Addison-Wesley Publishing Company.

Kelso, J.A. Scott. (1995). *Dynamic Patterns: The Self-Organization of Brain and Behavior*. Cambridge: MIT Press.

Kontopoulos, K. (1993). *The Logics of Social Structure*. New York: Cambridge University Press.

Lewin, R. (1992). *Complexity: Life at the Edge of Chaos*. New York: Macmillan.

Lorenz, E. (1993). *The Essence of Chaos*. Seattle: University of Washington Press.

Mainzer, Klaus. (1994). *Thinking in Complexity*. New York: Springer-Verlag.

Morgan, G. (1986). *Images of Organization*. London: Sage Publications.

Pickover, C. (ed.) (1997). *Fractal Horizons: The Future Use of Fractals*. New York: St. Martin's Press.

Priesmeyer, H. Richard. (1993). *Organizations and Chaos*. New York: Productivity Press.

Prigogine, I. and Stengers, I. (1988). *Order Out of Chaos*. New York: Bantam New Age Books.

Scott, G. P. (Ed.). (1991). *Time, Rhythms, and Chaos in the New Dialogue with Nature*. Ames, Iowa: The Iowa State University Press.

Stacey, Ralph. (1993). *Managing the Unknowable*. New York: Jossey Bass.

Stacey, Ralph. (1996). *Creativity and Complexity in Organizations*. San Francisco: Berrett-Koehler Publishers.

Talbot, M. (1991). *The Holographic Universe*. New York: Harper Perennial.

Waldrop, M. M. (1992). *Complexity: The Emerging Science at the Edge of Order and Chaos*. New York: Simon & Schuster.

Weick, K. (1979). *The Social Psychology of Organizing*. New York: Random House.

Wheatley, Margaret J. (1992). *Leadership and the New Science*. San Francisco: Berrett-Koehler Publishers.

Order Form

Coping with Chaos
Seven Simple Tools

Does your world get more complicated every day? Have your old rules and solutions lost their power in this fast-paced, chaotic world? *Coping with Chaos* can help. In her practical and easy-to-read manner, Glenda Eoyang introduces chaos theory and its implications for working productively in organizations. *Coping with Chaos* is full of thoughtful insights and helpful suggestions for coping with the chaos of everyday change. Order your copy today!

Please send me *Coping with Chaos*

_____ copies @ $19.95* each, total........ _____

Shipping & handling, $3.00 for the first
     book, $1.00 per book for add'l
     copies... _____

Subtotal.. _____

Sales tax (WY only) add 6%..................... _____

TOTAL*   _____

*U.S. dollars only, please.** For Canadian orders please include $28.00 per copy along with $4.50 shipping & handling for the first book, $1.50 for each additional book and include 7% GST.
  Bulk discounts are available, please write or fax for details.
    For Visa & MasterCard orders, please complete section below:
Name _____
Card Number _____ Expiration _____
Signature _____
Please mail (including full payment) or fax (credit card orders only) to:

Lagumo Corp.
1914 Thomes Avenue
Cheyenne, Wyoming 82001

Fax: 307-635-0365